THE AI UFOLOGIST

Answering the Big Questions of Ufology

RYAN S. WOOD

Preface

What are the big questions issues confronting ufology and the nature of the extraterrestrial presence? Questions like what is the ET agenda? or Why are they abducting humans? Do they care about humans? Why are they not helping us become a civilized global democracy?

These were the questions that inspired me to write, edit this book and then there were deeper and more sophisticated questions that intersect the challenges and opportunities within Ufology, government, and society writ large.

Fundamental to these questions is the concept that major governments have covered up the true nature of the UFO phenomenon for more than nearly a century through unprecedented secrecy, deception, and intimidation. This book provides a wealth of insights, although generated mostly by artificial intelligence they should astound and surprise even the most learned ufologist. Besides the answers to these questions, let me introduce a new ufology research and investigation tool—UFOdex.

UFOdex.com is two research and investigation tools: an artificial intelligence portal and a search portal into a massive silo of UFO data. Enabled by the fusion of several emerging information technologies: in artificial intelligence (AI), search, scanning, text analysis, comprehension, and automated deep web search tools, it promises to bridge the gap between simple web searches and true information intelligence.

The goal of UFOdex is the acquisition and integration of all UFO knowledge throughout time. All the national archives official documents, books, periodicals, credible papers, all the videos, images, in one comprehensive fully indexed silo of UFO intelligence. In short, our mission is UFO disclosure through intelligence dominance. That is the goal, practically it will take time and money to achieve the results.

Yet, UFOdex is more than just AI and Google on steroids, it is relevant data to the search topic at hand—it is encyclopedic knowledge, insights, and correlation in ultimately a multimedia format. UFOdex, when fully implemented will allow users to see at-a-glance the search results, but filter them by topics, such as relevant people, companies, events, and other hidden information contained within the knowledge base. Potentially, more important is the promising trajectory for artificial intelligence to glean the intelligence that the UFO/UAP researcher is looking for and be able to cross-reference that with high quality sources.

We are at a critical time in the history of the UFO enigma, a cusp if you will, where disclosure seems possible, but out of reach. To believe the government and military will reveal their most hidden super-secret for the sake of uplifting humanity, our energy independence and our future is, well, naive. So, we must take matters into our own hands, we must show leadership and take action to dismantle the UFO cover-up, step by step, year by year, until the evidence is overwhelming to the public, the media, our elected officials, and the keepers of the technological secrets.

The UFOdex tool is necessary to challenge the skeptical and conservative media, scientists and public. It will in time provide scholarly referenced answers for inquiring minds, thus elevating the UFO/UAP debate beyond the common hoax and speculative dismissals of swamp gas, Venus, and quacks. It will also become an engine for correlation, research, publication, and investigation that is unequaled in UFO history.

The AI Ufologist is the first pass at using UFOdex to answer the big questions of Ufology. The read will be provocative, insightful, amazing and at times possibly overwhelming.

Ryan S. Wood
Broomfield, Colorado
November 2023

Introduction

The pursuit of understanding the unknown has always been a fundamental drive of human curiosity, and nothing epitomizes the unknown more than the mysteries surrounding Unidentified Flying Objects (UFOs) and the potential of extraterrestrial life. "The AI Ufologist" embarks on an ambitious journey to bridge the gap between the enigmatic world of UFOs and the rapidly evolving field of artificial intelligence (AI). At the heart of this pioneering work lies a profound question: Can AI unravel the deepest mysteries of ufology that have eluded human understanding for decades, if not centuries?

For years, ufology has grappled with more questions than answers. What is the true nature of the extraterrestrial presence? Why might otherworldly beings be visiting Earth, and what are their intentions? Theories abound, ranging from scientific curiosity to benevolent guardianship, and from indifferent observation to potentially malevolent agendas. In a field crowded with speculation, anecdotal evidence, and unverified claims, this book introduces a groundbreaking tool – UFOdex. UFOdex.com represents a paradigm shift, harnessing the power of AI to sift through a massive trove of UFO data, encompassing national archives, official FBI documents, UFO books and periodicals, credible reports, and multimedia sources.

The core philosophy driving this exploration is the belief that AI, with its unparalleled data processing and pattern recognition capabilities, can offer new insights and perspectives on the UFO phenomenon. By transcending human biases and limitations, AI promises to bring a level of analytical rigor and massive depth that traditional ufological research has at times struggled to achieve. The objective of UFOdex is nothing short of comprehensive: to amass, integrate, and analyze all available UFO/UAP knowledge. This audacious mission aims to achieve UFO disclosure through intelligence dominance, a goal both ambitious and fraught with challenges. It will take time to capture all the unique and rare data that is not in the public domain.

As we delve into the capabilities and potential of UFOdex, we confront critical questions at the intersection of ufology, government secrecy, and societal implications. How might AI redefine our search for extraterrestrial intelligence? What can we learn from an AI-driven analysis of reported abductions, sightings, and government documents? Moreover, how can AI help navigate the complex web of government cover-ups and public misinformation that has long shrouded UFO studies?

"The AI Ufologist" starts this adventure by targeting some of the most common questions about the nature of UFOs, and their occupants. The book is not just a testament to technological advancement; it's a journey into the heart of human curiosity and the relentless quest for truth. It reflects a critical moment in history where the lines between science fiction and reality blur, offering a glimpse into a future where AI might hold the key to understanding our place in the cosmos. This book does not claim to have all the answers, there are always more thoughtful questions. Instead, it proposes a new path, an inspired blueprint to seeking truths and transitioning our appreciation and understating of this enigmatic and very often confusing, complex, and complicated field that has long been clouded by mystery and skepticism.

We stand at a pivotal moment where disclosure seems tantalizingly close yet remains elusive. In the quest to dismantle the decades-long UFO cover-up, UFOdex emerges as a beacon of hope – a tool that might finally tilt the balance in favor of transparency and truth. This book is a call to arms for ufologists, technologists, academics, and the curious public. It challenges us to rethink our approach to the UFO enigma and embrace the potential of AI as a tool for discovery and enlightenment.

As you turn these pages, prepare to embark on a journey that weaves together the threads of advanced technology, hidden knowledge, and the perennial human quest to understand the unknown. "the AI Ufologist" is more than a book; it's a portal into a future where the mysteries of UFOs may finally be unraveled, not by human minds alone, but through the lens of artificial intelligence and the actions of the people that choose to act for a better world.

Table of Contents

of testimony, insights, and reports about the phenomenon. Could this facilitate or accelerate the motivation for governments to admit the ET reality? ***page 102***

30. What are the five main forces that are interconnected driving global challenges? How are government debt and money economic forces affecting this? How is the political and societal polarization of the left and right mindset and actions influencing our future? Compare and contrast the geopolitical attitudes, policies, and priorities of countries particularly US and China when it comes to the planets future evolution? Addressing climate changes and mitigation costs will be a major global expense, how should equitable cost sharing be devised? Technology progress in energy, artificial intelligence, and biological sciences such as CRISPR, are they key enablers to a more peaceful planet with less war, inhumanity, and poverty? ***page 105***

1

What is the ET Agenda?

If extraterrestrial beings are visiting Earth, what are their intentions? Are they benign, malevolent, or indifferent to human affairs?

1. **Scientific Study**: Extraterrestrials may view Earth as a rich laboratory for scientific investigation. Like human scientists who meticulously study wildlife, these beings might be conducting extensive research on Earth's diverse ecosystems, human behavior, and biological peculiarities. Their agenda could primarily focus on gathering data, understanding the natural world, and possibly comparing it with other planetary systems.

2. **Resource Acquisition**: Extraterrestrials might have an interest in Earth's abundant natural resources. This could range from biological resources, like unique genetic material found in our diverse flora and fauna, to physical resources such as minerals, water, and rare earth elements. Additionally, they might be attracted to Earth's rich cultural heritage and knowledge, aiming to understand or even assimilate these assets into their own civilization.

3. **Benevolent Guidance**: Some theories propose that extraterrestrials have a protective, guiding agenda towards humanity. They might aim to assist in our evolution, steering us away from self-destructive paths and towards a more sustainable future. They could act as cosmic mentors, subtly influencing or directly intervening to prevent global catastrophes.

4. **Malevolent Intentions**: In stark contrast, there are theories suggesting that extraterrestrials could have harmful intentions towards humanity. Their agenda might include exploitation of humans and Earth's resources, enslavement of the human population, or even plans to completely eradicate humanity to repurpose the planet for their own needs.

5. **Indifference**: It's plausible that extraterrestrials could be utterly indifferent to human existence. Their operations on Earth or in its vicinity might be for reasons completely unrelated to humanity, such as conducting research on Earth's geology or atmosphere, or using the planet as a waypoint for interstellar travel, with little concern for human activities.

6. **Survival**: Extraterrestrials could be in search of a new home due to the loss or in hospitability of their own planet. Earth might present itself as a viable option for colonization or temporary refuge, prompting these beings to explore and assess the planet for its suitability for their survival needs.

7. **Diplomatic Relations**: They might be interested in establishing formal communication and diplomatic ties with humanity. This agenda could stem from a desire to integrate Earth into a broader interstellar community, sharing knowledge, culture, and technology as part of a cosmic federation or alliance.

8. **Observation**: Extraterrestrials might be conducting a long-term observational study of human civilization. This could involve monitoring the development of societal structures, technological advancements, and cultural evolution, as part of a larger study on the development and behavior of sentient life forms in the universe.

9. **Experimentation**: There are theories that Earth and its inhabitants are part of a vast extraterrestrial experiment. This could range from genetic experiments on humans and other Earth species to social and environmental manipulation, possibly to observe outcomes or develop new technologies.

10. **Spiritual or Metaphysical Goals**: Extraterrestrials might pursue objectives that are beyond human comprehension, rooted in spiritual or metaphysical realms. Their actions could be part of an effort to elevate cosmic consciousness, influence spiritual evolution, or manipulate energy fields that are unknown to human science.

11. **Genetic Manipulation**: Extraterrestrials might be engaging in complex genetic manipulation with humans. This could involve using humans as vessels for creating hybrid species, possibly to enhance the genetic robustness of both species or to achieve other unknown biological goals.

12. **Salvation of the Human Species**: Aliens may be acting as guardians of humanity, concerned about our future survival. This could involve plans to rescue a select group of humans in the event of a global catastrophe, ensuring the continuation of the species.

13. **Preservation of Genetic Stocks**: Extraterrestrials could be actively involved in archiving Earth's biological diversity. They might be collecting and preserving DNA samples from Earth's vegetation and animal species, possibly to safeguard against planetary disasters or for use in terraforming other worlds.

14. **Cosmic Consciousness**: Aliens might represent an advanced cosmic consciousness that seeks to infuse human thought with higher knowledge and awareness. Their agenda could be to gradually guide humanity towards a deeper understanding of the universe, expanding our consciousness and enlightening us about our place in the cosmos.

2

Why Would Extraterrestrials Abduct Humans?

Assuming the reports of abductions are accurate, what is the purpose behind these abductions? Is it for research, experimentation, or something else?

1. **Biological and Genetic Research**: A prevalent theory suggests that extraterrestrials are engaged in complex biological or genetic research, involving the study of human physiology and genetic makeup. This research could be driven by various motivations, such as understanding the human body's adaptability and resilience, exploring genetic diversity for scientific purposes, or even attempting to create hybrid species. The latter could stem from a need to preserve their own species or as part of an experiment in creating life forms suited to different environments. The idea of creating hybrids might also be linked to exploring evolutionary potentials or understanding how different species' genetics can be integrated.

2. **Neurological and Psychological Studies**: Extraterrestrials might be fascinated by the human brain and consciousness, leading them to abduct humans for detailed studies. This could involve exploring the intricacies of human psychology, cognition, emotions, and the neurological system. Such studies might aim to understand how humans process information, react to stimuli, or how emotions are chemically and neurologically triggered. By studying the human mind, aliens could be trying to understand our species' unique traits, such as creativity, empathy, and the capacity for complex thought and language.

3. **Monitoring of Human Development**: Theories suggesting extraterrestrials are monitoring human evolution and development could imply a long-term observational study, akin to a longitudinal research project. They could be tracking how humans have evolved over centuries, both biologically and socially. This might include studying changes in physical attributes, intelligence levels, social structures, and cultural developments. Such long-term monitoring could provide aliens with insights into the process of evolution and societal advancement in a relatively young species like humans.

4. **Cultural Examination**: Aliens abducting humans to learn about our cultures might be conducting an anthropological study of human societies. This could involve understanding human languages, social interactions, art forms, religious beliefs, and cultural practices. By examining these aspects, extraterrestrials could be trying to comprehend the diversity of human cultures, how they evolve, and how cultural elements influence human behavior and societal organization. This form of study could be analogous to how human anthropologists study various cultures to understand the broader aspects of humanity.

5. **Spiritual or Metaphysical Reasons**: Some abduction narratives include elements of spiritual or metaphysical experiences, where abductees report undergoing a transformation in their consciousness or spiritual understanding. In these accounts, the abduction is portrayed not just as a physical experience but also as a journey of spiritual enlightenment. This could suggest that extraterrestrials are interested in the spiritual or metaphysical aspects of human beings, possibly exploring concepts like the soul, consciousness beyond physical form, or the potential for humans to experience a broader cosmic awareness.

6. **Technological Testing**: The use of advanced technology in abduction experiences could indicate that extraterrestrials are testing their devices or procedures on humans. This might be for medical research, understanding how human bodies interact with alien technology, or studying human responses to unfamiliar technologies. The technology could range from medical instruments for physical examination to

devices that interact with the human mind, inducing virtual experiences or manipulating memories.

7. **Preparation for Future Events**: The idea that aliens are preparing humans for impending events on Earth, such as environmental disasters or interstellar conflicts, introduces a more proactive role for extraterrestrials. This could involve imparting knowledge, altering human genetics to withstand future conditions, or even mentally conditioning humans to face upcoming challenges. Such preparation could be seen as a form of guidance or protection, possibly out of a sense of responsibility or shared destiny in the larger cosmos.

8. **Societal Manipulation**: Some more speculative theories propose that abductions are part of a broader strategy to manipulate human society. This could include instilling fear as a form of control, altering perceptions and beliefs to guide the course of human development, or subtly influencing key individuals to affect societal decisions. These conspiracy theories often suggest that extraterrestrials have a vested interest in the direction of human civilization, whether for benign or malevolent purposes, and use abductions to exert influence.

3

Do Extraterrestrials Care About Humans?

Is there any evidence to suggest that extraterrestrial beings have an interest in the welfare or future of humanity?

1. **Benevolent Guardians**: This viewpoint suggests that extraterrestrials, if they are visiting Earth, act as protectors or guardians of humanity. People who subscribe to this belief often report encounters or abductions where they receive messages about maintaining peace, protecting the environment, and fostering human spiritual and technological growth. These extraterrestrials are perceived as mentors or elder siblings, guiding humanity towards a more enlightened and sustainable existence, often warning against the dangers of nuclear weapons, environmental destruction, and human conflict.

2. **Indifferent Observers**: In this perspective, extraterrestrials are seen as neutral entities who observe humans much like a biologist studies animals, without any emotional investment. They might be conducting a vast anthropological study, cataloging various intelligent species they encounter across galaxies. Humanity, in this view, is just one of many civilizations they monitor, possibly as part of a broader understanding of life in the universe. Their interaction or lack thereof is driven by scientific interest rather than any moral consideration about the welfare of humanity.

3. **Malevolent Conquerors**: Contrasting with benevolent perspectives, some theories and fictional narratives depict extraterrestrials as hostile

beings with imperialistic aims. These extraterrestrials could see Earth as a resource to be exploited, whether for its biological diversity, mineral wealth, or strategic location. In these scenarios, humans are often viewed as obstacles, subjects for experimentation, or even as a resource in themselves. This view is fueled by narratives of alien invasions and abductions, where the intent is harmful or oppressive towards human beings.

4. **Curious Scientists**: In this view, extraterrestrials are akin to researchers or scientists, with Earth and humans as subjects of their study. Their interest in humanity is purely academic, driven by a desire to understand different life forms and cultures. They might study human physiology, society, culture, and technology to gain insights into how intelligent life evolves and functions in different environments. This scientific curiosity is devoid of emotional or moral considerations, focused solely on gathering knowledge and data.

5. **No Interest**: It is plausible that extraterrestrials, assuming they exist, might not yet be aware of humanity's existence or have consciously chosen not to engage with us. Reasons for this could vary widely: from a lack of interest in a species they deem too primitive, to a strict policy of non-interference in the development of other civilizations. This perspective suggests that extraterrestrials might adhere to a cosmic principle like the 'Prime Directive' in 'Star Trek', avoiding any interaction that could influence or alter a civilization's natural development.

6. **Galactic Diplomats**: Some believe extraterrestrials could be interested in humanity as potential members of a larger interstellar community. In this scenario, extraterrestrials are observing and possibly even guiding humanity towards a future where we are ready to interact with other intelligent species. This would involve a gradual introduction to the concepts of cosmic citizenship and interstellar diplomacy.

7. **Spiritual Guides**: There are theories suggesting that extraterrestrials have a spiritual interest in humans. This viewpoint posits that aliens are aiding in the evolution of human consciousness or helping

humanity reach a higher spiritual plane. Abductee narratives often include elements of profound spiritual awakening or enlightenment, suggesting extraterrestrials' interest is not just physical or scientific, but also spiritual.

8. **Historical Custodians**: Another perspective is that extraterrestrials might see themselves as custodians or record-keepers of the universe, with humanity being one of many civilizations they have chronicled over eons. In this view, they might be interested in the cultural, historical, and social developments of human societies, documenting the rise and fall of civilizations much like historians or archivists.

Each of these perspectives offers a different view on the potential nature of extraterrestrial interest in humanity, ranging from guardianship and guidance to indifference, scientific curiosity, or even potential hostility. These views are shaped by a combination of personal experiences, cultural narratives, and speculative theories.

4

Why Haven't Extraterrestrials Made Open Contact?

The question of why extraterrestrials, assuming they exist and are aware of us, have not made open contact with humanity is a matter of great interest and is central to the field of astrobiology and the study of potential extraterrestrial intelligence. Here are several hypotheses that have been proposed to explain this:

1. **The Rare Earth Hypothesis**: This hypothesis contends that Earth-like planets hosting complex life forms are exceedingly scarce, and intelligent life is even scarcer. Given the vastness of the universe, it's proposed that extraterrestrial civilizations might be so rare or distant that contact is highly improbable. This rarity could be due to a unique combination of factors and circumstances that Earth possesses, which are not commonly replicated on other planets or solar systems. If intelligent extraterrestrial life is exceedingly rare, they may not exist within a reachable distance, or their technological capabilities might not be advanced enough for interstellar travel or communication.

2. **The Zoo Hypothesis**: This hypothesis likens humanity to animals in a zoo, where extraterrestrial civilizations are aware of us but deliberately avoid contact to allow natural progression. This non-interference might be to study human development in an unaltered state, like how biologists study animals in a natural habitat without intervening. Extraterrestrials might be observing us from a distance, ensuring that

their presence remains undetected to avoid influencing human culture, social evolution, or technological development.

3. **The Prime Directive Hypothesis**: Inspired by the non-interference principle in "Star Trek," this hypothesis posits that advanced extraterrestrial civilizations might have ethical guidelines preventing them from contacting less advanced societies. This self-imposed directive could be to prevent cultural contamination, societal disruption, or the collapse of civilizations not ready for such an encounter. The idea is that premature contact could lead to dependency, loss of cultural identity, or even the collapse of the less advanced society due to shock or upheaval.

4. **The Great Filter**: The Great Filter hypothesis suggests that at some point in the evolutionary process, there is a stage so difficult to surpass that most life forms never make it through. This filter could be the emergence of life, the leap to multi-cellular organisms, the development of intelligence, or the capacity for advanced civilization. If humanity has already passed this filter, it might explain the lack of contact with other civilizations, as they might have been unable to overcome this critical step. Alternatively, if the filter lies ahead of us, it could mean that civilizations tend to self-destruct at a certain point of technological advancement.

5. **Lack of Interest or Need**: It's conceivable that extraterrestrial civilizations, should they exist, might find nothing of interest or value in human civilization. They might have already explored what we are just beginning to understand, or they may have evolved beyond material needs and desires, making our resources, technology, and knowledge unappealing. Their motivations, priorities, and interests could be so different from ours that they see no reason to engage with humanity.

6. **Communication Issues**: The difficulty in establishing communication with extraterrestrial civilizations might be far greater than we can currently comprehend. Differences in language, signs, and cognition could make communication nearly impossible. Additionally, the vast

distances of space might render communication impractical, as the time taken for messages to travel across light-years could make timely or meaningful dialogue unfeasible. The methods and technologies for communication might also be so advanced or different that we are unable to recognize or understand them.

7. **They Have, But We Don't Recognize It**: This hypothesis suggests that extraterrestrials might have already tried to communicate with us, but their methods are beyond our understanding or detection. Their signals might be in a form that we cannot decipher, or we may lack the technological capability to detect them. It's also possible that evidence of their attempts has been misinterpreted as natural phenomena or remains undiscovered.

8. **Non-Existence in Our Dimension**: Some speculative theories propose that extraterrestrial life might exist in dimensions or frequencies beyond human perception. This multi-dimensional existence would make it impossible for us to detect them with our current technology or senses. They might be coexisting with us, but in a state that is completely out of the realm of our understanding and sensory experience.

9. **Self-Destruction**: This hypothesis posits that intelligent species tend to self-destruct due to internal or external factors before or shortly after developing interstellar communication or travel capabilities. Factors like war, environmental degradation, resource depletion, or other self-inflicted catastrophes could lead to the collapse of civilizations. If this pattern is common in the universe, it might explain the lack of contact with other intelligent beings.

10. **They Are Here Unobserved**: Some people believe that extraterrestrials might already be among us, observing or interacting with humanity in ways that are not overtly detectable. They could be using advanced cloaking or stealth technologies that render them invisible or undetectable to human senses and current scientific instruments. This theory suggests that they might be studying us covertly, either out of scientific interest or for reasons unknown.

11. **The Fermi Paradox**: Named after physicist Enrico Fermi, this paradox highlights the contradiction between the high probability of extraterrestrial civilizations' existence and the lack of evidence or contact with such civilizations. The paradox raises questions about the reasons behind this apparent silence, despite the vast number of stars and potentially habitable planets in the galaxy. It suggests that either our understanding of the likelihood of extraterrestrial life is incorrect, or there are factors at play preventing contact or detection.

Each of these hypotheses offers a different perspective on the reasons behind the lack of open contact with extraterrestrial civilizations, reflecting the complex interplay of scientific, technological, and philosophical considerations in this field.

5

What Could Governments Be Hiding About UFOs?

What information might governments possess about UFOs or extraterrestrials, and why would they choose to keep such information secret?

The idea that governments might be hiding information about UFOs (Unidentified Flying Objects), or extraterrestrial life is a common theme there are several reasons why governments might choose to withhold information about UFO sightings or encounters:

1. **National Security Concerns**: Governments may perceive unidentified aircraft as a significant national security threat. If these objects demonstrate advanced technology or capabilities beyond current human understanding, acknowledging their presence could expose a nation's defense weaknesses. Additionally, admitting to the inability to identify, track, or engage with these objects could undermine a government's perceived strength and deterrence capabilities, both domestically and internationally.

2. **Technological Secrecy and Advancement**: If a government has recovered or encountered technology from UFOs that is not understood or replicated by current human knowledge, keeping this information secret becomes crucial. This secrecy is not just to prevent other nations from gaining similar technological insights, especially if these technologies could be weaponized, but also to allow the country in possession of such technology to potentially develop a significant

technological edge in various fields, including military, energy, and communications.

3. **Preventing Public Panic and Societal Disruption**: The confirmation of extraterrestrial life or advanced technology could lead to widespread public fear, panic, or even societal upheaval. Governments might prefer to maintain secrecy to avoid dealing with the unpredictable consequences of such a revelation, which could range from religious and philosophical upheaval to widespread anxiety about potential extraterrestrial threats.

4. **Control of Information and Narrative**: By controlling the flow of information about UFOs, governments can shape the narrative around these phenomena. This control allows them to manage public perception, prevent speculation from becoming rampant, and ensure that any potential strategic or diplomatic actions taken in response to these sightings are not compromised by public debate or panic.

5. **Uncertainty and Lack of Explanation**: Governments might be reluctant to admit their encounters with UFOs due to the lack of concrete explanations. Acknowledging these phenomena without providing answers could undermine public trust in government competency. It also exposes a vulnerability in terms of national defense and scientific understanding, which could be seen as a weakness.

6. **Preventing the Spread of Misinformation and Speculation**: By keeping tight control over UFO-related information, governments might aim to prevent the spread of conspiracy theories, misinformation, or unfounded speculation that could lead to public hysteria, distract from important policy issues, or even spark false alarms and panic.

7. **Complexities in International Relations**: The discovery or interaction with extraterrestrial life or technology could drastically alter the global geopolitical landscape. Governments might withhold such information to carefully strategize on how to handle these changes in international relations, including potential shifts in alliances, the

creation of new treaties, or the development of joint global strategies for dealing with extraterrestrial matters.

8. **Uninterrupted Scientific Research**: Governments might prefer to conduct scientific studies on UFOs away from the public eye to ensure that the research is thorough, unbiased, and free from external pressure or influence. This secrecy allows scientists to methodically study these phenomena without the distraction of media sensationalism or public pressure, potentially leading to more accurate and considered conclusions.

9. **Impact on Economic Stability**: The revelation of advanced extraterrestrial technology could have unpredictable effects on global markets and industries. For instance, if this technology includes new energy sources or materials, it could render existing industries obsolete, leading to economic instability. Governments might choose to withhold this information to prevent market panic or to develop strategies to integrate these technologies in a more controlled and gradual manner.

10. **Legal and Ethical Challenges**: Encounters with extraterrestrial entities or technology would raise numerous legal and ethical questions. Issues could range from the rights of extraterrestrial beings (if they are sentient) to the legal status of their technology. Governments might be hesitant to disclose such encounters until they have developed a framework to address these complex legal and ethical dilemmas.

In recent years, some governments have become more open about their investigations into UFOs. For example, the U.S. Department of Defense has declassified several videos captured by Navy pilots, and the U.S. government has released reports acknowledging the presence of aerial phenomena that they cannot currently explain.

6

How Would Confirmation of Extraterrestrial Life Affect Human Society?

How might our societies, religions, and philosophies change if the existence of extraterrestrial life were to be confirmed? The confirmation of extraterrestrial life would likely have profound implications for human society, affecting a wide range of fields from science to theology. Here are some potential impacts:

1. **Revolution in Scientific Fields**: The confirmation of extraterrestrial life would not just be a milestone for biology; it would revolutionize numerous scientific disciplines. Astronomers, astrobiologists, chemists, and physicists would delve into understanding the nature of life beyond Earth. This could lead to new theories about the origins of life, the conditions needed for its existence, and the possibility of life in environments previously considered uninhabitable.

2. **Profound Shifts in Philosophy and Theology**: The discovery would prompt major philosophical and theological debates. Religions might undergo a period of introspection and reinterpretation of doctrines. Philosophers would grapple with questions about human uniqueness, the nature of consciousness, and our place in the universe. This could lead to a broader, more inclusive view of life and existence, transcending traditional human-centered perspectives.

3. **Intense Existential Inquiry**: On a personal level, individuals might experience profound existential contemplation. Questions

about human purpose, the uniqueness of human consciousness, and our responsibilities in a universe where we are not alone would become central. This could lead to a significant shift in personal values and life goals, as people reassess their place in a larger cosmic context.

4. **Technological Advancements and Economic Shifts**: The discovery could accelerate technological development, especially if extraterrestrial life possesses advanced technology. This could lead to new industries and economic shifts, potentially disrupting current market dynamics. The race to understand and possibly integrate alien technology could create new economic leaders and decline in traditional industries.

5. **Global Unity or Division**: The global reaction could go in two distinct directions. On the one hand, it could foster a sense of unity among humans, as we collectively face the reality of extraterrestrial life. On the other hand, it could exacerbate divisions, with different cultural, national, and religious groups reacting in disparate ways, potentially leading to conflict over how to engage with or interpret the new reality.

6. **Reconfiguration of Global Politics**: The discovery would necessitate international cooperation in formulating policies for space exploration, defense, and potential contact. It could either strain international relations, as nations compete for access to new knowledge and technology or strengthen them through collaborative efforts in facing a common extraterrestrial reality.

7. **Cultural Renaissance and Creative Inspiration**: The arts would likely experience a renaissance as humanity seeks to express and process this discovery. New genres of literature, film, and music might emerge, exploring themes of extraterrestrial life and its implications. This cultural shift could play a crucial role in shaping public understanding and acceptance of extraterrestrial existence.

8. **Expansion of Legal and Ethical Frameworks**: The presence of extraterrestrial life would raise complex legal and ethical questions. Laws might be developed to govern interactions with extraterrestrial beings or to manage the use of extraterrestrial resources. Ethical considerations, such as the rights of non-human life forms and the moral implications of space colonization, would become pressing issues.

9. **Transformation in Education**: Educational curricula would be revised to include the existence of extraterrestrial life. This could spark a surge in interest in STEM fields, particularly in astrobiology, astronomy, and environmental science. The educational focus might shift towards a more universal perspective on life and science.

10. **Psychological and Emotional Responses**: The societal psychological impact could range from excitement and wonder to fear and apprehension. Mental health professionals might need to develop new frameworks to help people process and adapt to this new reality. The way individuals and societies cope with this knowledge could vary widely, depending on cultural, psychological, and environmental factors.

11. **Enhanced Environmental Awareness**: The discovery could lead to a renewed emphasis on preserving Earth's biodiversity. Recognizing the fragility and rarity of life in the universe might inspire more concerted efforts to protect our planet's ecosystems and address environmental challenges.

The magnitude and nature of these impacts would indeed vary greatly depending on the specifics of the discovery. The detection of simple microbial life on another planet, for instance, would likely have a different societal impact compared to contact with an intelligent extraterrestrial civilization. Each scenario presents its own set of challenges, opportunities, and profound questions for humanity to explore.

7

What Are the Implications for National Security of Public ET Confirmation?

How do sightings of unexplained aerial phenomena pose a risk to national security, and how should governments respond to potential threats?

Sightings of unexplained aerial phenomena (UAP), particularly those that cannot be immediately identified as conventional aircraft or natural occurrences, can have several implications for national security:

1. **Enhanced Airspace Security Concerns**: The presence of UAP in a nation's airspace could signify a serious breach of territorial integrity. This raises critical questions about a country's ability to protect and monitor its airspace, especially if these objects demonstrate capabilities beyond current technological understanding. It also challenges existing protocols for airspace management and defense readiness.

2. **Implications of Superior Technology**: The possibility that UAPs represent technologically advanced systems, potentially far surpassing our own, poses a significant strategic threat. It implies that an adversary, terrestrial or otherwise, might possess capabilities that could undermine national defense systems, requiring a reevaluation of current military strategies and technologies.

3. **Challenges in Rapid Response and Military Readiness**: The inability to swiftly identify and react to UAPs highlights potential vulnerabilities in national defense mechanisms. This situation stresses the need for advanced detection systems and rapid response protocols to

address unknown aerial threats effectively, ensuring national security is not compromised.

4. **Need for Improved Interagency Communication**: The emergence of UAP necessitates robust communication channels among military branches, intelligence communities, and civilian aerospace authorities. Effective information sharing and coordination are crucial for assessing potential threats and formulating a unified response.

5. **Determining the Nature and Intent of UAP**: The ambiguous nature of UAPs makes it challenging to ascertain their purpose or intent. Governments must develop strategies to determine whether these phenomena represent a hostile presence, a peaceful or neutral entity, or something that requires diplomatic outreach.

6. **Opportunities and Risks in Technological Acquisition**: If a UAP is captured and confirmed to be of extraterrestrial origin, the opportunity to study and possibly reverse engineer its technology could lead to unprecedented technological leaps. However, this also raises concerns about the ethical implications and the potential for an arms race or power imbalances at the global level.

7. **Addressing Public Safety and Civil Aviation Concerns**: The appearance of UAPs could pose risks to civilian aviation and public safety, particularly if these objects enter heavily trafficked airspaces or are mistaken for hostile entities. Ensuring public safety requires clear protocols for civilian aviation encounters with UAPs and effective public communication strategies.

8. **Securing Sensitive Information**: The risk of leaks or cyberattacks targeting sensitive data related to UAP sightings and responses is a significant concern. Strengthening cybersecurity measures and protocols for handling classified information related to UAPs is essential to maintain national security integrity.

9. **Navigating International Diplomacy and Cooperation**: The global nature of the UAP phenomenon may require international

collaboration for effective monitoring, research, and response. This collaboration could be complex, given varying national interests, levels of technological advancement, and political agendas.

10. **Legal and Ethical Framework Development**: Engaging with UAPs, particularly if extraterrestrial in origin, introduces a host of legal and ethical dilemmas. These include establishing protocols for engagement, use of force, and addressing the rights or status of non-human entities. Developing international laws and ethical guidelines would be crucial.

In terms of how governments should respond to potential threats posed by UAP, the following steps are often suggested:

1. **Investment in Advanced Monitoring Technologies**:

    - **Development of Specialized Equipment**: Investing in the development of cutting-edge radar systems, infrared cameras, and other surveillance technologies that can detect and track UAPs with greater accuracy is crucial. This could involve funding research into new types of sensors that can capture a broader range of data, including unconventional flight patterns or energy emissions that are not characteristic of known aircraft or natural phenomena.
    - **Integration of AI and Machine Learning**: Utilizing artificial intelligence and machine learning algorithms can significantly enhance the analysis of data collected on UAPs. These technologies can help in pattern recognition, anomaly detection, and predictive analytics, providing deeper insights into UAP behavior and characteristics.
    - **Satellite Surveillance Enhancement**: Expanding the use of satellite technology to monitor UAP activities from space can provide a global perspective, offering real-time data and broader coverage areas that ground-based systems might miss.

2. **Comprehensive Interagency and International Collaboration**:

- **Establishing Protocols for Information Sharing**: Creating standardized protocols for sharing UAP-related information among various national agencies, including defense, aviation, and scientific bodies, is essential. This ensures that critical data is not siloed and can be rapidly disseminated for analysis and response.
- **International Data Exchange Platforms**: Developing international platforms for data exchange and collaboration can facilitate a more comprehensive understanding of UAPs. This could involve setting up secure communication channels and databases accessible to participating nations and relevant organizations.
- **Joint Research and Development Initiatives**: Encouraging joint research initiatives between countries can pool resources and expertise, leading to more effective monitoring systems and analytical tools. This collaboration can also help in standardizing how UAP incidents are reported and analyzed globally.

3. **Balanced Public Communication Strategy**:

- **Regular Updates and Transparency**: Providing regular updates to the public about UAP findings and investigations can help maintain transparency. This approach can mitigate the spread of misinformation and prevent conspiracy theories from taking root.
- **Educational Campaigns**: Implementing educational campaigns to inform the public about UAPs, their potential implications, and the steps being taken to understand them can help demystify the subject and reduce fear or speculation.
- **Crisis Communication Planning**: Preparing how to communicate in scenarios where a UAP poses a potential threat is important. This involves developing clear messaging that can quickly inform and guide the public without causing panic.

4. **Incorporation of Scientific Expertise**:

- **Partnerships with Academic Institutions**: Forming partnerships with universities and research institutions can bring a wealth of scientific knowledge to the study of UAPs. These collaborations can lead to innovative approaches and methodologies for investigation.

- **Multidisciplinary Research Teams**: Assembling teams that include astronomers, physicists, engineers, and other specialists can provide a more holistic understanding of UAPs. This diversity in expertise can help in deciphering complex data and formulating theories based on scientific principles.

5. **Ongoing Risk and Threat Assessment**:

   - **Continuous Monitoring and Analysis**: Establishing systems for the continuous monitoring of UAPs ensures that any potential threats are identified early. Regular analysis of this data helps in understanding any changes in patterns or behaviors of UAPs.
   - **Scenario Planning and Simulations**: Conducting simulations and developing various scenarios, from benign encounters to hostile engagements, can prepare governments for different possibilities. This planning should include potential diplomatic, military, and civil responses.

6. **Establishment of International Norms**:

   - **Global Treaties and Agreements**: Working towards global treaties and agreements on UAPs can help establish international norms and guidelines for engagement, response, and research. This could involve agreements on sharing data, joint responses to threats, and collaborative scientific studies.
   - **Standardization of Response Protocols**: Developing standardized response protocols for UAP sightings and encounters ensures a coordinated and effective global response. This includes protocols for military engagement, communication strategies, and scientific investigation.

7. **Training and Preparedness Programs**:

- **Specialized Training for Military and Aviation Personnel**: Implementing training programs for military pilots, air traffic controllers, and other aviation personnel on how to identify, report, and respond to UAP sightings is crucial. This training should include protocols for documenting encounters and safety measures.
- **Emergency Response Preparedness**: Preparing emergency response teams for scenarios involving UAPs, including potential crash retrievals or encounters, ensures readiness for any situation. This includes training in containment, public safety measures, and coordination with military and scientific teams.
- **Public Awareness and Education**: Educating the public on how to report UAP sightings and what to do in the event of an encounter can be valuable. This could involve community programs and collaborations with local authorities to disseminate information effectively.

The approach to UFO/UAP and national security is evolving, especially as the topic gains more attention and as governments take the phenomena more seriously, as evidenced by official investigations and reports released by various countries, including the United States.

8

How Should Humanity Respond to a Confirmed Extraterrestrial Threat?

If extraterrestrial activity poses a clear and present danger, what would be the appropriate global response? Here are some steps and considerations that might be involved in formulating an appropriate response:

1. **Global Unity and Cooperation**:

    - **Formation of a Unified Global Coalition**: This would involve not just the United Nations, but also various international organizations, scientific communities, and even private sector stakeholders, coming together to form a comprehensive response strategy.
    - **Shared Intelligence and Resources**: Ensuring that intelligence, resources, and research are shared among nations would be critical. This would help in pooling global expertise and capabilities, essential for an effective response.

2. **Scientific Assessment**:

    - **International Scientific Task Forces**: Assembling teams of top scientists from various disciplines (astrophysics, biology, technology) to analyze the threat comprehensively.
    - **Understanding Extraterrestrial Technology and Biology**: Efforts to understand the science behind any encountered

extraterrestrial technology or biology would be crucial in developing effective countermeasures or communication strategies.

3. **Risk Communication**:

   - **Regular Updates and Information Channels**: Establishing dedicated channels for disseminating accurate information to the public, possibly through a centralized international body or coordinated media outlets.
   - **Managing Public Sentiment**: Implementing strategies to manage public sentiment and prevent hysteria, including community engagement programs and educational campaigns.

4. **Defense Strategies**:

   - **Global Defense Collaboration**: Collaborating on a global scale to develop defense strategies, which could include space-based defense systems or ground-based tactical responses.
   - **Innovation in Defense Technologies**: Fast-tracking the development of new defense technologies, potentially harnessing insights from the scientific assessment of the extraterrestrial entities.

5. **Diplomatic Efforts**:

   - **Establishment of Extraterrestrial Communication Protocols**: Developing protocols and strategies for attempting communication with the extraterrestrial entities, with the aim of understanding their intentions and seeking a peaceful resolution.
   - **International Diplomatic Channels**: Utilizing existing international diplomatic channels or creating new ones specifically for extraterrestrial engagement.

6. **Evacuation and Contingency Planning**:

   - **Global Evacuation Strategies**: Developing international plans for the evacuation of populations from high-risk areas, including logistics for transportation, accommodation, and sustenance.

- **Safe Zones and International Support**: Establishing international safe zones and ensuring support for displaced individuals, including medical care, food, and shelter.

7. **Economic Measures**:

   - **Stabilization Policies and Emergency Funds**: Implementing economic stabilization policies and setting up emergency funds to support affected industries and maintain market confidence.
   - **Resource Allocation and Trade Adjustments**: Adjusting trade policies and resource allocation to ensure the availability of critical supplies and services.

8. **Legal Framework**:

   - **Development of Extraterrestrial Engagement Laws**: Crafting new international laws or amending existing ones to address the unique legal challenges posed by extraterrestrial engagement.
   - **Human Rights and International Law**: Ensuring that responses adhere to international human rights standards and the principles of international law.

9. **Cultural and Ethical Considerations**:

   - **Inclusive Decision-Making**: Ensuring that the cultural diversity of humanity is represented in decision-making processes.
   - **Ethical Guidelines for Extraterrestrial Interaction**: Developing ethical guidelines for any form of interaction with extraterrestrial entities, respecting both human values and potential extraterrestrial sensibilities.

10. **Preservation of Knowledge**:

   - **Digital and Physical Repositories**: Creating repositories of human knowledge, both digital and physical, that could survive catastrophic events. This could include backups of scientific, cultural, and historical data.

- **Global Heritage Sites**: Designating and protecting sites of significant cultural and historical importance as part of humanity's collective heritage.

11. **Continuity of Government and Society**:

    - **Emergency Government Protocols**: Establishing protocols to ensure the continuity of government functions under extraterrestrial threat scenarios.
    - **Community Leadership and Local Governance**: Strengthening local governance and community leadership to maintain order and provide basic services during crises.

12. **Psychological Support**:

    - **Mental Health Services and Counseling**: Providing widespread access to mental health services and counseling for populations dealing with the stress and anxiety of the extraterrestrial threat.
    - **Community Support Systems**: Building strong community support systems to help maintain social cohesion and address collective psychological impacts.

13. **Research and Development**:

    - **Targeted R&D Initiatives**: Directing research and development efforts towards specific areas that could counter the extraterrestrial threat, such as advanced propulsion systems, energy shields, or communication technologies.
    - **Collaborative Research Programs**: Establishing international collaborative research programs to leverage global expertise and resources.

14. **Non-Military Solutions**:

    - **Environmental and Technological Mitigation Strategies**: Exploring environmental modification or advanced technological solutions that could mitigate the threat without resorting to conflict.

- **Peaceful Coexistence Strategies**: Developing strategies for peaceful coexistence in case the extraterrestrial entities are not hostile but simply present in our environment.

15. Post-Threat Reconstruction:

- **Rebuilding and Rehabilitation Plans**: Developing comprehensive plans for the reconstruction and rehabilitation of affected areas post-threat, including infrastructure rebuilding, economic revitalization, and community support.
- **Lessons Learned and Future Preparedness**: Analyzing the response to the extraterrestrial threat to learn lessons and improve future preparedness, including revising international protocols and response strategies.

The hope would be that through cooperation, innovation, and peaceful negotiation, humanity could avert or mitigate the threat. However, the specific response would greatly depend on the details of the threat itself—its nature, immediacy, and the options available to humanity at the time.

9

What is the Role of Private Companies in Ufology and Space Exploration?

With private companies increasingly involved in space exploration, what role might they play in uncovering evidence of extraterrestrial life, and how transparent would they be?

The involvement of private companies in space exploration has been growing, and their role in ufology and the search for extraterrestrial life is becoming more significant. Here's how they might contribute and the issues surrounding transparency:

1. **Innovation and Technology Development**:

    - **Cutting-Edge Instruments and Tools**: Private companies are developing sophisticated tools and instruments that could detect bio-signatures or techno-signatures, indicators of life or technology used by extraterrestrial beings.
    - **Revolutionizing Space Travel**: Innovations in spacecraft design, life support systems, and sustainable habitats for long-duration missions could open new possibilities for deeper space exploration and the search for life.

2. **Space Missions**:

    - **Targeted Exploration Missions**: Beyond Mars, companies could target Europa, Enceladus, and other celestial bodies with potential for harboring life, using advanced probes and landers.

- **Private Rovers and Drones**: The deployment of private rovers and drones on other planets or moons could significantly increase the scope of exploration and the chances of finding evidence of life.

3. **Data Collection and Analysis**:

   - **Advanced Data Processing**: With access to advanced computing resources, private companies could process and analyze space data more efficiently, potentially identifying signs of life faster than traditional methods.
   - **Sharing Data with Scientific Community**: Establishing protocols for data sharing with the global scientific community could enhance the collective understanding of potential extraterrestrial life.

4. **Partnerships with Government Agencies**:

   - **Joint Missions and Research Projects**: Collaborative projects with agencies like NASA or ESA could combine government experience with private sector innovation, leading to more comprehensive exploration missions.
   - **Resource and Information Sharing**: These partnerships could facilitate the sharing of resources, expertise, and information, accelerating the pace of discovery.

5. **Funding and Resources**:

   - **Private Funding for Ambitious Projects**: The influx of private capital can fund ambitious projects that might be beyond the scope of government budgets, such as missions to distant exoplanets or the development of new exploration technologies.
   - **Diversifying the Search for Life**: Private funding could support a wider range of projects, from traditional scientific research to more unconventional methods of searching for extraterrestrial intelligence.

6. **Public Engagement**:

- **Inspiring Public Interest**: High-profile missions and charismatic business leaders can generate public excitement and interest in space exploration and the search for extraterrestrial life.
- **Educational Outreach**: Private companies could engage in educational programs and public outreach, fostering a new generation of scientists and enthusiasts.

7. **Transparency Issues**:

- **Balancing Commercial Interests with Public Knowledge**: The tension between protecting intellectual property and contributing to scientific knowledge could impact the level of transparency.
- **Selective Disclosure**: Companies may choose to disclose only certain findings that align with their business interests or public image, potentially skewing public understanding.

8. **Regulation and Oversight**:

- **Developing New Space Laws**: As private companies become more involved in space, there may be a need for new laws and regulations to govern their activities, especially those that could impact the search for extraterrestrial life.
- **Preventing Contamination**: Ensuring that private missions comply with planetary protection protocols to prevent biological contamination is crucial.

9. **Ethical and Legal Considerations**:

- **Debates Over Information Ownership**: Questions about who owns the rights to extraterrestrial discoveries and how this information should be used could become increasingly complex.
- **Updating International Treaties**: The Outer Space Treaty and other agreements may need revisions to address the specific challenges and scenarios posed by private space exploration.

10. Potential for New Discoveries:

- **High-Risk, High-Reward Missions**: Private companies might be more inclined to invest in high-risk missions, exploring new frontiers in the search for extraterrestrial life.
- **Innovative Approaches to SETI**: Investment in projects like SETI or new methods of scanning exoplanets could yield results that traditional space programs have not achieved.

In summary, private companies have the potential to greatly advance the field of ufology and the search for extraterrestrial life, but their role also introduces complexities regarding transparency, data sharing, and the commercialization of space. As these companies become more prominent in space exploration, there will likely be an ongoing dialogue about the best ways to balance private enterprise with public interest and international standards.

10

How Can Public Interest and Government Cover-up Policy Be Reconciled?

Given the public's demand for transparency and the government's need for secrecy in certain areas, how can a balance be struck that satisfies both parties?

1. **Clear Communication:**
   - **Detailed Explanations for Secrecy:** When withholding information, providing detailed explanations for the need for secrecy can help in gaining public understanding and trust.
   - **Regular Public Briefings:** Holding regular briefings on national security matters, even if some details are withheld, can keep the public informed and involved.

2. **Legislation and Oversight:**
   - **Strengthening Legal Frameworks:** Strengthening the legal frameworks that govern classified information can help prevent abuses of power and ensure that secrecy is justified.
   - **Enhanced Oversight Mechanisms:** Establishing or enhancing oversight mechanisms, such as parliamentary committees or independent watchdogs, can provide checks and balances on government secrecy.

3. **Declassification Processes:**

- **Timely Review of Classified Materials**: Implementing a system for the timely review and declassification of materials can ensure that information is not kept secret longer than necessary.
- **Public Access to Declassified Documents**: Making declassified documents easily accessible to the public, for example through online archives, can increase transparency.

4. **Public Engagement**:

- **Public Consultations and Forums**: Organizing public consultations and forums on national security policies can provide a platform for citizen input and concerns.
- **Involvement in Policy Development**: Involving the public or their representatives in the development of policies related to national security and secrecy can enhance democratic participation.

5. **Whistleblower Protections**:

- **Robust Legal Protections for Whistleblowers**: Ensuring strong legal protections for individuals who expose wrongdoing is crucial for maintaining governmental integrity.
- **Safe Channels for Reporting Misconduct**: Creating safe and confidential channels for whistleblowers to report misconduct can encourage the exposure of abuses while protecting sensitive information.

6. **Independent Media**:

- **Supporting Investigative Journalism**: Encouraging and supporting investigative journalism can play a significant role in uncovering and reporting on government activities.
- **Media Literacy Programs**: Promoting media literacy among the public can help people critically evaluate news sources and understand the role of the media in a democracy.

7. **Educational Initiatives**:

- **National Security Education Programs**: Implementing educational programs that explain the complexities of national security and the reasons for certain secrecy measures.
- **Public Workshops and Seminars**: Organizing workshops and seminars on topics related to state secrets and public transparency can educate citizens on these issues.

8. **Transparency in Non-Sensitive Areas**:

- **Open Government Initiatives**: Launching open government initiatives where non-sensitive information is made readily available to the public.
- **Transparency Reports**: Publishing regular transparency reports on government activities that do not compromise national security.

9. **International Collaboration**:

- **Global Standards for Secrecy and Transparency**: Working towards global standards and best practices for balancing secrecy and transparency in government affairs.
- **International Oversight Bodies**: Participating in or supporting international oversight bodies that monitor and report on state secrecy and transparency.

10. **Privacy Protections**:

- **Balancing Public Interest with Privacy Rights**: Implementing policies that carefully balance the need for transparency with respect for individual privacy rights.
- **Data Protection Regulations**: Enforcing strict data protection regulations to ensure that personal information is not unnecessarily exposed in the pursuit of transparency.

11. **Regular Review and Adaptation**:

- **Dynamic Policy Adjustment**: Ensuring that policies on secrecy and transparency are not static but are regularly reviewed and adapted to changing circumstances and public needs.
- **Feedback Mechanisms**: Establishing mechanisms for public and expert feedback on the effectiveness and appropriateness of secrecy policies.

12. Public Advocacy and Civil Society:

- **Empowering Advocacy Groups**: Supporting and empowering advocacy groups and NGOs that focus on government transparency and accountability.
- **Partnerships Between Government and Civil Society**: Fostering partnerships between government agencies and civil society organizations to work collaboratively on issues of public interest and transparency.

Achieving the right balance between transparency and secrecy is an ongoing process that must adapt to changing circumstances and public sentiment. It requires vigilance from both the government and the public to ensure that the scales do not tip too far in either direction. These questions reflect the broad spectrum of concerns and curiosities that ufology encompasses. They touch upon scientific inquiry, philosophical considerations, policy decisions, and the profound implications that any confirmed extraterrestrial presence would have on humanity. The search for answers continues to inspire both scientific and popular interest, and the debate over disclosure versus secrecy remains a contentious issue in the field.

11

Why is the government hiding the truth about extraterrestrial visitation and crashed UFOs?

1. **Strategic Advantage in Technology**:
    - **Military Applications**: The potential military applications of extraterrestrial technology could be revolutionary, offering capabilities far beyond current human technology. Governments might seek to develop these applications in secret to maintain a military edge.
    - **Technological Monopoly**: By keeping alien technology, a secret, a government could enjoy a monopoly on its use and development, potentially dominating key sectors like energy, aerospace, and materials science.

2. **Prevention of Panic and Social Unrest**:
    - **Psychological Impact on Society**: The revelation of extraterrestrial life could profoundly impact the collective psyche, potentially leading to fear, denial, or even societal upheaval.
    - **Historical Precedents**: Governments might draw from historical instances where new scientific discoveries or encounters with foreign civilizations caused significant disruption and apply this caution to the disclosure of extraterrestrial life.

3. **National Security Concerns**:

- **Vulnerability to Extraterrestrial Threats**: Acknowledging extraterrestrial visitations could imply that a government is vulnerable to threats beyond its control, potentially undermining national security.
- **Surveillance and Reconnaissance**: If extraterrestrial entities have superior surveillance capabilities, governments might want to keep their knowledge and countermeasures secret.

4. **Control Over Advanced Technologies**:

- **Economic Advantage**: Control over alien technology could provide immense economic advantages, potentially revolutionizing industries and creating new markets.
- **Research and Development**: Governments might restrict access to alien technology to conduct research and development in controlled environments, free from external pressures or theft.

5. **Avoiding Public Distrust and Speculation**:

- **Government Competence**: Revealing the existence of extraterrestrial visitations might lead to public questioning of government competence, especially if it appears that the government has been unable to protect or understand these phenomena.
- **Conspiracy Theories**: Ironically, by keeping information secret, governments may inadvertently fuel conspiracy theories and public speculation about their activities and intentions.

6. **International Relations and Diplomacy**:

- **Information Sharing Among Nations**: The decision to share or withhold information about extraterrestrial encounters could significantly impact international trust and cooperation.
- **Geopolitical Power Dynamics**: The possession of extraterrestrial technology could drastically alter geopolitical power dynamics, potentially leading to conflicts or shifts in alliances.

7. **Preservation of Religious and Cultural Norms**:

- **Impact on Religious Beliefs**: The existence of extraterrestrial life could challenge fundamental religious doctrines, potentially leading to a crisis of faith for many.
- **Cultural Identity and Narratives**: Societal narratives and cultural identities that have been built around human uniqueness and the understanding of life could be fundamentally altered.

8. **Scientific Uncertainty and Caution**:

- **Incomplete Understanding**: Governments might be cautious about releasing information on phenomena that are not yet fully understood, to avoid misinformation or premature conclusions.
- **Responsible Disclosure**: There may be a desire to ensure that any disclosure is backed by solid scientific evidence and understanding, to provide accurate and responsible information to the public.

9. **Limiting Economic Disruption**:

- **Impact on Traditional Industries**: The introduction of advanced alien technology could render many existing technologies and industries obsolete, causing significant economic disruption and job losses.
- **Market Stability**: Sudden revelations about alien technology could lead to market instability, with unpredictable effects on global economies.

10. **Collaboration with Alien Entities**:

- **Secret Agreements or Communications**: There might be agreements or communications with extraterrestrial beings that require confidentiality, either for security reasons or as part of the terms of the interaction.
- **Protecting Extraterrestrial Privacy**: Governments might be acting to protect the privacy and wishes of extraterrestrial entities who may not desire public disclosure.

In summary, the reasons for government secrecy regarding extraterrestrial visitations and UFOs could range from concerns over national security and technological control to the desire to prevent societal upheaval and maintain economic stability. The complexity of these issues means that any disclosure would need to be handled with great care and consideration for the potential global impact.

What are some of the reasons why the government should admit the truth about UFOs?

1. **Promoting Transparency and Trust**:
    - **Restoring Faith in Public Institutions**: In an era marked by skepticism towards authority, transparent communication about UFOs could restore public faith in governmental institutions.
    - **Setting a Precedent for Openness**: Acknowledging UFOs could set a new standard for openness in government, encouraging transparency in other contentious or mysterious areas.

2. **Alleviating Public Fear and Misinformation**:
    - **Educational Outreach**: Official disclosure could be accompanied by educational campaigns to inform the public, replacing fear with understanding and curiosity.
    - **Combating Conspiracy Theories**: Transparency could dispel many of the conspiracy theories that thrive in the absence of reliable information, thereby reducing misinformation.

3. **Validating Eyewitness Accounts**:
    - **Psychological Relief**: Official recognition of UFO sightings would offer psychological relief to witnesses, acknowledging their experiences as legitimate.

- **Encouraging Future Reporting**: This validation could encourage others to come forward with their sightings, contributing to a more comprehensive understanding of the phenomena.

4. **Addressing the Weight of Evidence**:

    - **Scientific Integrity**: Acknowledging the evidence would demonstrate a commitment to scientific integrity and empirical understanding.
    - **Respect for Research Efforts**: It would honor the work of researchers and investigators who have dedicated years to studying the phenomena, often in the face of skepticism and ridicule.

5. **Ensuring National Security**:

    - **Improved Defense Mechanisms**: Understanding UFOs is essential for national defense; official recognition could lead to improved mechanisms to identify and respond to aerial threats.
    - **Global Security Collaboration**: It could foster international collaboration on potential aerial threats, enhancing global security.

6. **Encouraging Scientific Inquiry and Research**:

    - **New Fields of Study**: Government admission could lead to the development of new fields of study, potentially revolutionizing our understanding of physics, aerospace, and other sciences.
    - **Funding and Resources**: It could open up new avenues for funding and resources dedicated to UFO research, attracting top scientists and thinkers to the field.

7. **International Cooperation and Transparency**:

    - **Global Standards for UFO Reporting**: This could lead to the establishment of global standards and protocols for reporting and investigating UFO incidents.
    - **Cross-Cultural Exchange of Information**: It would encourage a cross-cultural exchange of information and experiences related to UFOs, enriching our collective understanding.

8. **Confronting the Cover-Up Allegations**:

   - **Restoring Public Confidence**: Addressing these allegations directly could restore public confidence in the government's willingness to deal honestly with its citizens.
   - **Legal and Ethical Accountability**: It would demonstrate a commitment to legal and ethical standards in government operations.

9. **Legal and Ethical Responsibility**:

   - **Informed Consent**: Keeping citizens informed is a cornerstone of democratic societies, allowing for informed public discourse and consent.
   - **Preparation for Future Scenarios**: Disclosure would help prepare the public for any future scenarios involving UFOs or extraterrestrial life, ensuring that society is not caught off-guard.

10. **Cultural and Societal Impact**:

    - **Inspiration for Science and Technology**: Acknowledging UFOs could inspire new generations in the fields of science, technology, and astronomy, much like the space race did in the mid-20th century.
    - **Philosophical and Existential Discussions**: It could spark profound philosophical and existential discussions about humanity's place in the universe and our understanding of life.

In summary, the government's admission of the truth about UFOs could have far-reaching implications, fostering a culture of transparency and trust, advancing scientific knowledge, ensuring national security, and profoundly impacting our cultural and societal understanding. This openness would not only validate the experiences and efforts of many but also prepare humanity for future encounters or discoveries in this enigmatic field.

13

Explain why UFOs and ET's can or are a National Security Threat

The potential national security implications of UFOs and ETs can be understood from several perspectives:

1. **Unknown Capabilities**:

    - **Advanced Weaponry and Defense Systems**: ETs could possess weaponry and defense systems that render our current military capabilities obsolete, posing an immediate threat to global security.
    - **Surveillance and Intelligence Gathering**: If ETs have superior surveillance capabilities, they could gather sensitive information, compromising national security and privacy.

2. **Airspace Violations**:

    - **Challenges to Air Defense Systems**: Unidentified craft in a country's airspace challenge the effectiveness of existing air defense systems, potentially exposing vulnerabilities.
    - **Impact on Civil Aviation**: The presence of UFOs in commercial flight corridors could pose risks to civil aviation, necessitating new air traffic control protocols.

3. **Technological Advancement**:

   - **Uneven Technological Development**: If one nation gains access to ET technology, it could lead to uneven technological development, disrupting the global balance of power.
   - **Arms Race**: The pursuit of ET technology could spark a new kind of arms race, with nations vying for control over alien resources and knowledge.

4. **Panic and Social Unrest**:

   - **Challenges to Social Order**: The revelation of ET existence could undermine the social order, as traditional institutions and beliefs are questioned.
   - **Impact on Mental Health**: Widespread anxiety and uncertainty about ET intentions could have a significant impact on public mental health.

5. **Misidentification**:

   - **Accidental Military Engagements**: Misidentifying UFOs as hostile could lead to accidental military engagements, potentially triggering international conflicts.
   - **Strain on Military Resources**: Constant monitoring and identification efforts could strain military resources, diverting attention from other security concerns.

6. **Economic Implications**:

   - **Disruption of Traditional Industries**: The introduction of advanced ET technology could disrupt traditional industries, leading to economic instability.
   - **New Economic Opportunities**: Conversely, ET technology could create new economic opportunities and industries, reshaping the global economy.

7. **Information Control**:

- **Challenges in Managing Public Perception**: Controlling the narrative around UFOs and ETs is crucial to prevent misinformation and manage public reaction.
- **Cybersecurity Concerns**: The need to protect sensitive information about ETs could heighten cybersecurity concerns, as leaks could have unpredictable consequences.

8. **Potential Biological Threat**:

    - **Risk of Pandemics**: ET organisms could carry pathogens that humans have no immunity against, posing a risk of new pandemics.
    - **Impact on Earth's Ecosystem**: Alien life forms could disrupt Earth's ecosystem if they are not contained properly.

9. **Distrust in Government**:

    - **Erosion of Public Confidence**: If governments are found to have concealed information about ETs, it could lead to a significant erosion of public trust in authorities.
    - **Political Repercussions**: The revelation of withheld information could have political repercussions, including protests and demands for greater transparency.

10. **Global Implications**:

    - **Need for International Collaboration**: The discovery of ET life would necessitate unprecedented international collaboration to address the shared challenge.
    - **Potential for Conflict or Unity**: The global response to ETs could either lead to increased international conflict or foster a sense of unity against a common external entity.

In summary, the potential national security implications of UFOs and ETs are vast and multifaceted, ranging from immediate threats to air safety and military security to broader societal and global challenges. These issues would require careful, coordinated responses from governments, international organizations, and communities worldwide.

14

Provide descriptions and theories about how UFOs and ET craft operate technologically

Based on various reports, claims, and theories, several hypotheses have been proposed about their potential technological operation:

1. **Operating Characteristics**:

    - **Adaptive Flight Behavior**: UFOs might possess advanced navigation systems that allow them to adapt to different environments, from space to Earth's atmosphere.
    - **Control Systems**: The control mechanisms of these craft could be far beyond physical controls, potentially involving advanced AI or even mind-machine interfaces.

2. **Physical Appearance**:

    - **Material Technology**: The materials used in UFO construction might have properties like self-healing, adaptability to different atmospheric conditions, or radiation shielding.
    - **Aerodynamic Structures**: Despite their unconventional shapes, these craft might be designed for optimal aerodynamics, possibly altering shape in response to different flight conditions.

3. **Speed and Sound**:

    - **Supersonic Capabilities**: UFOs might be capable of supersonic speeds without creating sonic booms, suggesting a mastery over air displacement and shock waves.

- **Vibration and Frequency Manipulation**: The occasional sounds associated with UFOs might be a byproduct of their propulsion systems, possibly involving manipulation of sound frequencies or vibrations.

4. **Propulsion Systems**:

   - **Advanced Energy Manipulation**: Beyond Element 115 claim of Bob Lazar, these systems might involve harnessing energy from zero point vacuum fluctuations or other exotic energy sources, such as fusion.
   - **Space-Time Manipulation**: Theories suggest that these craft might not just distort space/time for propulsion but could also create localized time distortions.

5. **Electro-Gravitic Propulsion**:

   - **Gravity Control**: This technology might involve creating artificial gravity wells, allowing the craft to "fall" in any direction at incredible speeds.
   - **Energy Field Propulsion**: The propulsion could be based on creating and manipulating energy fields that interact with gravitational forces.

6. **Anti-Gravity Propulsion**:

   - **Repulsion and Attraction Forces**: Anti-gravity might involve balancing repulsive and attractive gravitational forces, allowing for precise and agile movements.
   - **Gravitational Wave Manipulation**: Advanced understanding and manipulation of gravitational waves could be key to this technology.

7. **Advanced Propulsion Systems**:

   - **Plasma Propulsion**: Some speculate the use of highly energized plasma for propulsion, offering high thrust and efficiency.

- **Quantum Propulsion**: Theories include the use of quantum mechanics principles, like quantum tunneling, for instantaneous movement across distances.

8. **Warp Drives and Wormholes**:

   - **Exotic Matter Utilization**: These technologies might require exotic matter or negative energy to create stable wormholes or warp bubbles.
   - **Dimensional Folding**: The concept of folding space to achieve instantaneous travel between two points, akin to a cosmic shortcut.

9. **Inertial Dampening**:

   - **Inertia Negation**: Advanced inertial dampening systems might negate the effects of inertia, allowing occupants to withstand extreme maneuvers.
   - **Gravitational Field Manipulation**: Altering the craft's internal gravitational field to counteract external forces.

10. **Stealth and Cloaking**:

    - **Light Bending and Refraction**: Technologies that bend or refract light around the craft, making it invisible to the naked eye and radar.
    - **Quantum Stealth**: Theoretical quantum cloaking methods that could render an object undetectable across various spectrums.

11. **Energy Source**:

    - **Harnessing Cosmic or Dark Energy**: Utilizing abundant cosmic or dark energy as a power source.
    - **Advanced Nuclear Technologies**: Beyond fusion, these could include matter-antimatter annihilation or harnessing energy from black holes.

12. **Material and Construction**:

- **Nanotechnology and Self-Assembly**: Materials that can self-assemble or rearrange at a molecular level for different functions.
- **Environmental Adaptation**: Materials that adapt their properties based on external environmental conditions.

13. **Interaction with Consciousness**:

- **Telepathic Interfaces**: Technology that translates thoughts or emotions into commands or communications.
- **Consciousness-Driven Reality Manipulation**: The craft's technology might be intertwined with the consciousness of its operators, responding to their thoughts or intentions.

14. **Multidimensional Travel**:

- **Interdimensional Gateways**: The craft might be capable of opening gateways to different dimensions or parallel universes.
- **Reality Warping**: Advanced understanding of the fabric of reality could allow these craft to traverse or manipulate different dimensions.

15. **Biological Integration**:

- **Cybernetic Organisms**: Craft that are part biological, part machine, possibly with self-healing or adaptive capabilities.
- **Symbiotic Technology**: Technology that integrates or works in harmony with biological entities, possibly adapting to the needs or health of its occupants.

These theories and descriptions, while speculative, represent an attempt to understand and explain the advanced technologies that might be employed by UFOs and extraterrestrial craft, based on the observed phenomena, and reported encounters.

15

What are the functions of MJ-12 in relation to the extraterrestrial presence? What are their goals and objectives?

"Majestic 12" or "MJ-12" is the name of a secret committee of scientists, military leaders, and government officials formed in 1947 under an executive order by U.S. President Harry S. Truman. The group was created to facilitate recovery and investigation of alien spacecraft and to manage all information regarding extraterrestrial life and technology, following the purported UFO incident in Roswell, New Mexico.

The purported functions of MJ-12, according to these theories, include:

1. **Recovery Operations**:

    - **Incident Management**: MJ-12 is believed to be responsible for managing the logistics and operations surrounding the recovery of alien spacecraft, including securing crash sites and coordinating with military and intelligence agencies.
    - **Entity Retrieval and Care**: In cases where extraterrestrial entities are involved, MJ-12 might oversee their retrieval, medical care, and communication efforts, ensuring their well-being while maintaining secrecy.

2. **Scientific Research**:

    - **Interdisciplinary Studies**: The committee is thought to oversee research that spans various scientific disciplines, from physics and aerospace engineering to biology and linguistics, to understand alien technology and biology.
    - **Collaboration with Top Secret Labs**: MJ-12 may collaborate with highly classified laboratories and research facilities, both governmental and private, to study and reverse-engineer extraterrestrial artifacts.

3. **Information Control**:

    - **Media Manipulation and Disinformation**: Part of their role could involve controlling the narrative in the media, possibly disseminating disinformation to obscure the truth about extraterrestrial encounters.
    - **Document Classification and Secrecy**: Ensuring that all documents, reports, and findings related to extraterrestrial activities are highly classified and accessible only to authorized personnel.

4. **Policy Formulation**:

    - **Strategic Planning**: Developing long-term strategies for how humanity would deal with extraterrestrial life and potential threats or alliances.
    - **Legal Frameworks**: Creating legal frameworks and protocols for extraterrestrial engagement, including potential diplomatic or military interactions.

5. **Surveillance and Monitoring**:

    - **Global Monitoring Network**: Establishing a global network of surveillance to monitor UFO activity, employing advanced technologies for detection and tracking.

- **Counterintelligence Operations**: Conducting counterintelligence operations to prevent leaks and manage insider threats within organizations involved in UFO research and encounters.

6. **Collaboration with Other Nations and Entities**:

   - **International Agreements**: Forming covert agreements and collaborations with other nations that have knowledge of extraterrestrial activities, sharing information and resources.
   - **Engagement with International Organizations**: Possibly interacting with international organizations to establish a unified approach to extraterrestrial presence, while maintaining secrecy.

7. **Technological Application**:

   - **Military Advancements**: Utilizing extraterrestrial technology to develop advanced weaponry, defense systems, and strategic assets, potentially changing the balance of global military power.
   - **Aerospace and Energy Innovations**: Applying alien technology to advance aerospace capabilities, including propulsion systems, and possibly developing new energy sources.

8. **Preparation for Disclosure**:

   - **Public Relations Strategies**: Developing strategies for eventual public disclosure, including managing public reaction and integrating extraterrestrial knowledge into society.
   - **Scenario Planning**: Creating various scenarios and response plans for different types of potential public disclosures, ranging from gradual acknowledgment to emergency scenarios.

9. **Extraterrestrial Communication and Diplomacy**:

   - **Linguistic and Cultural Research**: Studying extraterrestrial languages and cultures to prepare for potential communication or diplomatic relations.

- **Negotiation and Liaison Roles**: Potentially acting as intermediaries in any direct communication or negotiations with extraterrestrial entities.

10. Ethical and Philosophical Considerations:

- **Ethical Frameworks**: Establishing ethical guidelines for interaction with extraterrestrial life, including treatment of extraterrestrial entities and use of their technology.
- **Philosophical and Societal Impact Studies**: Considering the broader philosophical and societal implications of extraterrestrial presence and integrating these considerations into policy and public communication strategies.

The goals and objectives of MJ-12 would likely be to protect national security interests, maintain geopolitical stability, advance technological capabilities, and manage societal impact related to extraterrestrial phenomena.

16

Assuming Extraterrestrial biology is unique and from a completely different evolutionary perspective, how and why would militaries weaponize captured ET's?

The hypothetical scenario of militaries attempting to weaponize extraterrestrial beings is a common theme in science fiction and speculative discussions about contact with alien life. The underlying assumption in these discussions is that extraterrestrial biology could offer unique advantages or properties that are not found in terrestrial organisms, which could be exploited for military purposes. Here are some reasons and methods by which militaries might seek to weaponize extraterrestrial life:

1. **Adaptive Camouflage**: Extraterrestrial organisms might possess natural forms of camouflage far superior to anything on Earth. This could be studied and replicated to develop advanced stealth technologies for military personnel or equipment.

2. **Enhanced Regeneration**: If extraterrestrials have rapid healing or regenerative abilities, this could be invaluable in medical treatment for soldiers, potentially leading to the development of new regenerative medicines or treatments.

3. **Extreme Physical Conditions**: ETs might survive in conditions that are lethal to humans (extreme temperatures, radiation, etc.). Understanding and replicating these traits could lead to enhanced

protective gear or vehicles capable of operating in otherwise inhospitable environments.

4. **Telepathic or Psychic Abilities**: If extraterrestrials possess telepathic or other psychic abilities, militaries might study these to develop new forms of communication, interrogation techniques, or even psychological warfare tactics.

5. **Advanced Propulsion or Locomotion**: Studying the movement mechanisms of extraterrestrial beings could lead to breakthroughs in propulsion or vehicle design, offering new strategies for troop movement or deployment.

6. **Superior Material Science**: The biological makeup of extraterrestrials might include materials with properties superior to any known-on Earth, such as extreme durability or the ability to self-repair. These materials could be replicated for use in armor, weaponry, or construction.

7. **Climate Adaptation**: If extraterrestrials are adapted to a wide range of climates, their biology could offer insights into creating gear or systems that allow soldiers to operate more effectively in diverse environmental conditions.

8. **Enhanced Cognitive Abilities:** If ETs have advanced problem-solving skills, memory, or other cognitive abilities, these traits could be studied to enhance the strategic and tactical capabilities of military personnel.

9. **Bioluminescence or Energy Emission**: Some extraterrestrial life forms might emit light or other forms of energy naturally. This could be harnessed for signaling, illumination, or even as a direct energy source for military applications.

10. **Symbiotic Relationships**: Exploring how extraterrestrials form symbiotic relationships with other organisms or their environment could lead to new strategies in warfare, such as creating symbiotic relationships between soldiers and engineered organisms or materials for enhanced capabilities.

In weaponizing extraterrestrial life, militaries would be venturing into uncharted ethical territory, raising significant questions about the treatment of sentient beings and the potential consequences of integrating alien biology into Earth's ecosystems. The pursuit of such advancements would surely be shrouded in secrecy, given the profound implications for global security and the balance of power. It's important to note that these ideas arise from a human-centric viewpoint that tends to project our own patterns of behavior, such as weaponization, onto unknown entities. The ethical, legal, and practical implications of encountering extraterrestrial life would likely be complex and multifaceted. Moreover, international treaties, such as the Outer Space Treaty, currently prohibit the weaponization of space and the use of celestial bodies for military purposes, which would include the weaponization of extraterrestrial life.

17

There have been many reports that ET communications with humans occur without words and conversation takes place in the minds of individuals. If mental or telepathic and psychic communications were possible to be trained technologically for most humans how would that affect the trajectory or progress of humanity or cultures? Is this a positive or negative recognizing that it might take ten or more years for this skill to evolve and mature.

The potential for technologically enhanced telepathic or psychic communication among humans is a profound concept that could significantly alter the trajectory of humanity and culture. The effects of such a development would be far-reaching and multifaceted. Here's a speculative look at some of the potential impacts:

Positive Impacts

1. **Crisis Management and Emergency Response**: In emergency situations, telepathic communication could enable faster and more efficient coordination among first responders, potentially saving lives by improving the speed and accuracy of information exchange.

2. **Deepening Personal Relationships**: Telepathy could allow for a deeper, more intimate level of understanding and connection between

individuals, potentially strengthening personal relationships and fostering a greater sense of empathy.

3. **Enhanced Diplomacy**: In international relations, telepathy could facilitate more transparent and effective diplomatic communications, potentially reducing misunderstandings and conflicts between nations.

4. **Streamlined Legal and Judicial Processes**: In legal settings, telepathic communication could offer a more direct way of understanding intentions and truths, potentially transforming how testimony and evidence are presented and interpreted.

5. **Revolutionizing Therapy and Mental Health**: Therapists and mental health professionals could use telepathy to gain better insights into the thoughts and feelings of their clients, leading to more effective treatment strategies.

6. **Spiritual and Philosophical Growth**: Telepathy could lead to new spiritual and philosophical movements, as people gain a more direct understanding of different perspectives and experiences, potentially leading to a more interconnected and compassionate society.

Negative Impacts

1. **Overdependence on Technology**: There could be an overreliance on telepathic communication, leading to a decline in traditional forms of communication and potentially diminishing interpersonal skills like empathy and patience.

2. **Identity and Individuality Issues**: With the blurring of personal thoughts and the communal mind, individuals might struggle with maintaining a sense of identity and individuality, leading to existential crises.

3. **Manipulation and Propaganda**: The ability to transmit thoughts directly could be exploited for propaganda or manipulation on a massive scale, influencing public opinion and behavior more directly than ever before.

4. **Ethical Dilemmas in Law Enforcement**: The use of telepathy in law enforcement could lead to ethical concerns, such as the potential for invading the privacy of thoughts or the misuse of telepathic interrogation techniques.

5. **Impact on Traditional Cultural Practices**: Many cultures that rely on traditional oral storytelling or other forms of communication might face challenges in preserving these practices in a world dominated by telepathic communication.

6. **Potential for Mental Overload**: The human brain might be overwhelmed by the constant influx of telepathic communication, leading to new types of cognitive disorders or a need for 'mental downtime' to prevent overload.

Evolution and Maturation of the Skill

As societies adapt to this new form of communication, there will likely be a period of significant social and cultural evolution. Educational systems might include training on how to use telepathic communication responsibly and effectively. Legal systems would need to evolve to address the new ethical and privacy challenges posed by telepathy. Additionally, there could be a development of new forms of etiquette and social norms to govern when and how telepathic communication is appropriate.

In summary, the advent of technologically enhanced telepathic communication would represent a monumental shift in human interaction, with profound implications across all aspects of society. Its successful integration would hinge on careful consideration of the ethical, social, and psychological impacts, ensuring that the benefits are maximized while minimizing potential harms.

18

Describe the nature, extent, and details of sexual encounters with aliens (extraterrestrials) and humans.

The topic of sexual encounters between humans and alleged extraterrestrial beings is a complex and sensitive one, primarily found within the realm of speculative narratives, personal claims, and the more controversial aspects of ufology. It's important to approach this subject with the understanding that there is no scientifically verified evidence to substantiate these claims, and they are often viewed with significant skepticism by the mainstream scientific community.

Nature of Claims

- **Abduction Narratives**: Some individuals who claim to have been abducted by aliens also report experiences that they interpret as sexual in nature. These narratives often involve medical examinations or procedures that are perceived as reproductive or sexual by the abductee.
- **Hybridization Programs**: A common theme in these accounts is the idea of extraterrestrial beings engaging in a form of hybridization, purportedly creating a mix of human and alien offspring. This is often described as part of a larger extraterrestrial agenda, the nature and purpose of which vary widely among different narratives.

Extent of Claims

- **Varied and Anecdotal**: The claims of sexual encounters are varied and predominantly anecdotal. They come from a small subset of individuals who report alien encounters or abductions.
- **Global Reports**: While not widespread, such reports have emerged from various parts of the world, suggesting a global but fragmented set of narratives.

Details of Encounters

- **Physical Descriptions**: Descriptions of these encounters often include specific details about the extraterrestrial beings involved, ranging from humanoid forms to more abstract or unusual appearances.
- **Emotional and Psychological Impact**: Individuals reporting these experiences often describe significant emotional and psychological impact, ranging from trauma and fear to, in some cases, a sense of connection or communication with the beings.
- **Lack of Physical Evidence**: A notable aspect of these claims is the general lack of physical evidence supporting the encounters, which contributes to the skepticism they receive from the scientific and academic communities.

Contextual Considerations

- **Cultural and Psychological Factors**: Researchers and psychologists have pointed out that these claims may be influenced by cultural narratives, psychological factors, or other non-extraterrestrial phenomena. Some suggest that such experiences might be a manifestation of psychological stress, sleep paralysis, or other cognitive processes.
- **Skepticism and Debate**: The scientific community largely views these claims with skepticism, considering them unverified and outside the realm of empirical science. They are often categorized

as part of the broader UFO and alien abduction phenomena, which are themselves subjects of debate and controversy.

- **Role in Ufology**: In the field of ufology, these claims are a controversial and often marginalized topic. While some ufologists consider them an important aspect of understanding potential extraterrestrial contact, others caution against the uncritical acceptance of such narratives.

In conclusion, while claims of sexual encounters between humans and extraterrestrials are a part of some alien abduction narratives, they remain unsubstantiated by empirical evidence and are approached with skepticism within the scientific community. They are often viewed through the lenses of psychology, cultural studies, and the broader phenomenon of UFO and alien abduction narratives.

19

Money and capital control the economics of most of the global civilization, the business transactions, commerce, products, and services. Describe the technologies or situations that would disrupt or re-wire this political, economic, and social paradigm that exists today in relation to a concrete disclosure of extraterrestrial presence with crashed UFO hardware and dead alien entities?

The disclosure of concrete evidence of extraterrestrial presence, such as crashed UFO hardware and non-human entities, would likely have profound and multifaceted impacts on our political, economic, and social paradigms. Here are several technologies and situations that could arise from such a disclosure and potentially disrupt or rewire the current systems:

1. **Communication Technologies**: If extraterrestrial entities possess advanced communication technologies, such as instantaneous, long-distance communication methods, it could revolutionize how we interact globally. This could disrupt traditional telecommunications industries and change the dynamics of international relations, making remote collaboration and diplomacy more immediate and effective.

2. **Environmental Technologies**: Alien technology might include advanced methods for managing and repairing ecological damage, which could radically alter industries related to environmental conservation and climate change mitigation. This could shift economic priorities

from industries that are harmful to the environment towards more sustainable practices.

3. **Space Exploration and Colonization**: Advanced extraterrestrial technology could accelerate human space exploration and colonization efforts. This would not only have profound implications for space-related industries but could also lead to the exploitation of resources from other planets and celestial bodies, potentially reshaping global economies.

4. **Education and Training**: The need to understand and utilize alien technologies would likely lead to a complete overhaul of educational systems, with a new focus on science, technology, engineering, and mathematics (STEM), and potentially new fields of study directly related to extraterrestrial science and technology.

5. **Cultural Exchange and Influence**: The discovery of extraterrestrial cultures could introduce humanity to entirely new forms of art, music, literature, and philosophy. This cultural exchange might inspire a global renaissance of creativity and thought, influencing everything from entertainment to lifestyle choices.

6. **Legal and Ethical Frameworks**: The presence of extraterrestrial entities and technology would necessitate the development of new legal and ethical frameworks, particularly in terms of intellectual property, interaction protocols, and the rights of non-human entities.

7. **Shift in Military Paradigms**: The military might focus on preparing for potential extraterrestrial threats or collaborations, leading to a shift in defense strategies, including the development of space-based defense systems and the retraining of military personnel in new forms of warfare or collaboration.

8. **Human Identity and Unity**: The knowledge that we are not alone in the universe could foster a sense of global unity, as humans perceive themselves more as a single species with common interests and challenges, rather than divided by national or cultural differences.

9. **Investment and Funding Shifts**: The economic landscape could see a shift in investment and funding, with significant resources being redirected towards research, development, and integration of extraterrestrial technologies, potentially leading to the decline of traditional industries and the rise of new sectors.

10. **New Philosophical Movements**: The revelation of extraterrestrial life could give rise to new philosophical movements that question our understanding of existence, consciousness, and the nature of the universe, potentially leading to a shift in societal values and priorities.

11. **Tourism and Exploration**: Public interest in extraterrestrial locations and cultures could create a new tourism industry, with people seeking to experience these new frontiers firsthand, leading to economic opportunities and challenges in managing interstellar travel and exploration.

12. **Interstellar Trade and Economy**: The possibility of trade with extraterrestrial civilizations, whether in knowledge, resources, or technology, could create an entirely new economic sector, with implications for global trade and economy.

The extent of the disruption would depend on the nature of the technologies, the conditions under which they are disclosed, and the ability of governments and institutions to manage the transition. It's also possible that the initial disruption would eventually lead to a new equilibrium, with societies adapting to and integrating the new technologies into a redefined economic and social order.

In conclusion, the disclosure of extraterrestrial presence and technology would likely be a catalyst for unprecedented change, challenging our current economic, political, and social structures. The transition period could be tumultuous, but it also holds the potential for significant advancements and a redefinition of humanity's role in the cosmos.

20

If the goal is to elevate all global humanity toward a more educated, more food, water, shelter, and energy secure place with the norms of the democracy, basic human rights, and the rule of law, how should it be done? What technologies are required? How can the ET reality play a role in this transformation? What is the nature of religions in this shift to a fairer world?

Elevating global humanity to a state of greater education, security, and democratic governance is a multifaceted challenge that requires a combination of technological innovation, policy reform, cultural shifts, and international cooperation. Here's a broad outline of how this might be approached, including the potential role of extraterrestrial reality and the impact on religion:

Technological Requirements

1. **Education Technology**:
    - Development of AI-driven educational platforms that adapt to individual learning styles.
    - Virtual reality (VR) and augmented reality (AR) applications for immersive learning experiences.
    - Global online platforms facilitating cross-cultural educational exchanges and collaborations.

2. **Agricultural Technology**:
   - Precision agriculture using drones and IoT sensors for efficient resource use.
   - Development of climate-resilient crops through genetic engineering.
   - Urban agriculture solutions, including hydroponics and aquaponics, to reduce food transportation costs and emissions.

3. **Water Purification and Management**:
   - Nanotechnology in water filtration for efficient removal of pollutants.
   - Large-scale rainwater harvesting and sustainable water management practices.
   - Technologies for efficient wastewater treatment and reuse.

4. **Renewable Energy Solutions**:
   - Advanced solar photovoltaic (PV) cells and wind turbines with higher efficiency.
   - Development of safe nuclear fusion technology as a potential clean energy source.
   - Battery technologies with higher energy densities for better energy storage.

5. **Healthcare Innovations**:
   - Wearable health monitoring devices and AI in diagnostics for early disease detection.
   - Gene editing technologies like CRISPR for treating genetic disorders.
   - Telehealth services for remote areas to improve healthcare accessibility.

6. **Sustainable Infrastructure**:
   - Eco-friendly building materials and green architecture designs.

- Smart city technologies for efficient resource management.
- Development of resilient infrastructure to withstand climate change impacts.

7. **Global Connectivity**:

- Satellite internet to provide high-speed internet access in remote areas.
- Blockchain technology for secure and transparent global transactions.
- Enhanced cybersecurity measures to protect global digital infrastructure.

Role of ET Reality in Transformation

1. **Technological Leapfrogging**:

- Potential access to advanced ET technologies could accelerate solutions in energy, medicine, and environmental protection.
- Collaboration with ET intelligence might offer new scientific methodologies or principles.

2. **Global Perspective Shift**:

- ET disclosure could foster a global identity, reducing terrestrial conflicts and promoting a united approach to global challenges.
- It might inspire a new wave of exploration and curiosity, driving innovation and scientific inquiry.

3. **Ethical and Philosophical Evolution**:

- The realization of ET existence could lead to a reevaluation of human ethics and values.
- It could stimulate philosophical debates about life, consciousness, and our place in the universe.

Evolving Role of Religions

1. **Guidance in Ethical Use of Technology**:

    - Religions could provide ethical frameworks for the responsible use of new technologies, especially those impacting life and the environment.
    - They could help in addressing moral dilemmas posed by advancements like genetic engineering and AI.

2. **Promoting Interfaith Harmony**:

    - Religions could play a key role in promoting peace and understanding among different faiths considering a common, broader cosmic existence.
    - They could facilitate dialogues on universal values and ethics in a more interconnected world.

3. **Reinterpretation and Integration**:

    - Religious doctrines might undergo reinterpretation to incorporate a new understanding of life in the universe.
    - This could lead to more inclusive and expansive spiritual perspectives.

Implementation Strategies for Global Upliftment

1. **Inclusive and Sustainable Policy Development**:

    - Policies focusing on inclusive growth, ensuring that technological advancements benefit all sections of society.
    - Emphasis on sustainable development goals (SDGs) in policy planning.

2. **Fostering Global and Local Partnerships**:

    - Strengthening international collaborations for technology exchange and joint problem-solving.
    - Encouraging local community participation in decision-making to ensure that solutions are culturally and contextually relevant.

3. **Cultural and Social Adaptation**:
   - Programs to prepare societies for rapid technological changes and their social implications.
   - Emphasizing cultural preservation and appreciation in the face of globalizing influences.

4. **Economic Systems Aligned with New Realities**:
   - Exploring alternative economic models like circular economies to reduce waste and promote sustainability.
   - Policies to address potential job displacements due to automation and AI.

5. **Empowerment through Education and Access**:
   - Ensuring universal access to quality education, including digital literacy.
   - Vocational training and skill development programs aligned with future job markets.

6. **Ethical Governance of Emerging Technologies**:
   - Establishing global standards for the ethical development and deployment of emerging technologies.
   - Creating regulatory frameworks to address issues like privacy, data security, and AI ethics.

7. **Promoting Cultural Exchange and Global Citizenship**:
   - Encouraging cultural exchange programs to foster global understanding and empathy.
   - Education curricula incorporating global citizenship education to prepare younger generations for a more interconnected world.

In summary, achieving a fairer world with widespread security and democratic norms is a complex goal that requires coordinated action across multiple fronts. While technology is a critical enabler, the ultimate success will depend on human values, choices, and the ability to work together towards a common future. The potential discovery or interaction with extraterrestrial intelligence could be a catalyst for change, offering new perspectives and technologies, but it would still be up to humanity to steer this change in a positive direction.

21

Describe the most powerful ten science and technological innovations of the future, say in the year 2100 that could have a profound uplifting of humanity to war-free, fairer, and more secure world for its citizens. Would science revolutions of practical immortality, anti-gravity, low-cost portable fusion, psychic skills, space exploration, and artificial intelligence.

Predicting the future of science and technology is inherently speculative, but based on current trends and theoretical possibilities, we can imagine a range of innovations that could profoundly uplift humanity by the year 2100. Here are ten potential developments that could contribute to a more peaceful, fair, and secure world:

1. **Advanced Renewable Energy Systems**: Beyond fusion, we might see innovations like highly efficient solar cells, capable of capturing a much larger spectrum of sunlight, or wireless energy transmission, enabling remote areas to access power easily. This democratization of energy would reduce geopolitical tensions over energy resources and enable sustainable development globally.

2. **Molecular Nanotechnology**: This technology could lead to self-repairing materials, reducing waste and the need for new resources. In medicine, it could enable targeted drug delivery or cellular repair, dramatically improving health outcomes. Economically, it could lead

to a shift from large-scale manufacturing to localized, on-demand production of goods.

3. **Practical Immortality**: Beyond extending life, this could involve enhancing human resilience to diseases and aging. Societies would need to address ethical questions around population control and resource allocation, but the potential for a vastly experienced and knowledgeable population could accelerate human progress.

4. **Artificial General Intelligence (AGI)**: AGI could manage complex systems like global logistics networks, optimize agricultural outputs, and even assist in governance, potentially reducing corruption and inefficiency. Its ability to process and analyze vast amounts of data could lead to unprecedented levels of understanding and problem-solving.

5. **Quantum Computing and Communication**: Quantum technologies could also enable simulations of complex systems like entire ecosystems or economies, allowing for better planning and management. Secure quantum communication could protect privacy and intellectual property, fostering a safer and more innovative global research environment.

6. **Space Colonization and Resource Utilization**: This could involve mining asteroids for rare minerals, reducing the environmental impact of mining on Earth. Colonizing other planets or moons would not only be a safety valve for Earth's growing population but also a catalyst for new forms of social and political organization.

7. **Anti-Gravity or Advanced Propulsion**: Such technologies could also revolutionize space travel, making interstellar exploration feasible. On Earth, they could lead to new forms of architecture and city planning, as buildings and infrastructure are no longer limited by traditional gravitational constraints.

8. **Enhanced Cognitive Abilities**: This could also involve collective intelligence systems where human minds are networked together,

leading to a form of 'hive mind' that could tackle complex problems through unparalleled collaborative thinking.

9. **Global Surveillance and Predictive Policing**: While raising significant privacy concerns, if implemented with strict ethical guidelines, such systems could greatly enhance emergency response times, disaster management, and even global health monitoring, predicting, and preventing outbreaks before they spread.

10. **Telepathy and Enhanced Communication**: This could extend to a universal translation system, breaking down language barriers and fostering a truly global culture. Enhanced empathy could lead to more compassionate and cooperative international relations, reducing misunderstandings and conflicts.

Each of these innovations carries its own set of challenges and ethical considerations. Their development and implementation would need to be guided by a global consensus, ensuring that they are used for the betterment of humanity as a whole and not just a privileged few. The potential involvement of extraterrestrial intelligence or technology, whether through direct interaction or the reverse-engineering of recovered UFO/UAP technology, could accelerate or significantly alter the trajectory of these developments. The integration of such advanced technologies would also necessitate a reevaluation of our current socio-political structures and a shift towards more inclusive and globally minded governance.

The realization of these technologies would require careful governance to ensure they are used ethically and equitably. The potential for misuse or the exacerbation of existing inequalities presents significant risks that must be managed through international cooperation and foresight.

Moreover, the societal impact of such technologies would be profound. They would challenge current economic models, social structures, and philosophical paradigms. The concept of work, the distribution of wealth, the nature of conflict, and the essence of human identity would all be subject to reevaluation.

In a world with such advanced technologies, science and ethics would need to advance together to ensure that these innovations lead to the uplift-ment of humanity rather than its detriment. The key would be in ensuring that the benefits are shared equitably, and that human rights, autonomy, and freedom are preserved and enhanced in the face of transformative technological change.

22

What can be or is the role of artificial intelligence in accelerating the disclosure of the extraterrestrial presence to humanity. What are the best ways of using AI to achieve this end?

The role of artificial intelligence in potentially accelerating the disclosure of extraterrestrial presence to humanity could manifest in several ways, primarily through the enhancement of research capabilities and the analysis of large datasets. Here are some of the best ways AI could be utilized towards this end:

1. **Advanced Data Analysis and Pattern Recognition**: AI can sift through petabytes of astronomical data, identifying anomalies that human researchers might miss. This includes analyzing electromagnetic signals for non-random patterns, scanning exoplanet atmospheres for signs of artificial alteration, or detecting unusual interstellar objects.

2. **Signal Decryption and Linguistic Analysis**: In the event of receiving a complex extraterrestrial signal, AI could be instrumental in decryption efforts. It could analyze the structure of the signal, compare it with vast linguistic databases, and even employ machine learning to hypothesize the rules of an entirely alien language.

3. **Enhanced Image and Video Analysis**: AI's ability to analyze visual data with increasing sophistication means it could play a crucial role in authenticating UFO footage. By filtering out noise, enhancing resolution, and applying advanced pattern recognition, AI could help

distinguish genuine phenomena from natural occurrences or man-made objects.

4. **Predictive Modeling and Scenario Simulation**: AI can simulate a wide range of extraterrestrial encounter scenarios based on varying parameters. This could help in understanding potential extraterrestrial technology, behavior, and the implications of their presence on Earth.

5. **Comprehensive Simulation Capabilities**: Beyond first contact scenarios, AI could simulate the societal, political, and global impacts of extraterrestrial disclosure. This would aid in preparing comprehensive response strategies.

6. **Facilitating Global Research Collaboration**: AI can act as a nexus for international scientific collaboration, integrating diverse research efforts into a cohesive whole. It can identify gaps in our current understanding and suggest new research directions, potentially leading to breakthroughs in the study of extraterrestrial life.

7. **Public Engagement and Education**: AI-driven platforms could be used to educate the public about extraterrestrial life, manage expectations, and debunk myths. Interactive AI systems could personalize learning experiences and answer questions, thereby reducing fear and misinformation.

8. **Real-Time Monitoring of Public Sentiment**: By analyzing social media and news, AI can gauge public reaction to developments in the field of extraterrestrial research. This can help governments and organizations to tailor their disclosure strategies in a way that is sensitive to public sentiment.

9. **Crowdsourcing Data Analysis**: AI can efficiently process data from global crowdsourcing initiatives, where the public submits reports and evidence of UFO sightings. This democratizes the search for extraterrestrial life and allows for a broader base of observation.

10. **Navigating Ethical and Philosophical Terrain**: AI could be used to model and analyze the ethical implications of extraterrestrial contact.

By processing vast amounts of philosophical and ethical texts, AI could help in formulating policies that are considerate of diverse perspectives and potential extraterrestrial ethics.

In summary, AI's role in the potential disclosure of extraterrestrial presence is not just limited to data analysis and scientific research. It extends to public engagement, ethical considerations, and the preparation of humanity for one of the most profound events in its history. The integration of AI in this domain could significantly enhance our readiness and response to the discovery of extraterrestrial life.

To effectively use AI in the quest for disclosure, it's important to ensure that the algorithms are transparent, and the data sources are credible. Misinformation or biased data could lead to false conclusions and undermine the credibility of any findings. Moreover, the ethical use of AI in this context is paramount to maintain trust and integrity in the process.

Ultimately, while AI can significantly aid in the search for extraterrestrial intelligence and the analysis of related data, the decision to disclose findings will likely remain a human one, influenced by social, political, and cultural factors. AI can provide tools and insights, but it is up to human institutions and leaders to determine how to use them in the context of potential extraterrestrial disclosure.

23

Describe the alleged messages that come from Crop Circles in detail with particular emphasis on the technological ones.

Crop circles have long fascinated both the public and researchers, with various interpretations and theories about their origins and meanings. Among these, some claim that crop circles are a form of communication from extraterrestrial beings, containing messages for humanity. It's crucial to note, however, that these interpretations are speculative and not supported by empirical scientific evidence. The alleged messages from crop circles, particularly those related to technology, can be categorized, and described as follows:

Alleged Messages in Crop Circles

1. **Mathematical and Geometric Information**:

    - Some crop circles exhibit complex geometric patterns and have been interpreted by enthusiasts as containing advanced mathematical concepts, like fractals or sacred geometry.
    - These patterns are sometimes seen as demonstrating a higher understanding of mathematics and geometry that could have potential applications in technology.

2. **Astronomical and Astrological Symbols**:
    - Certain crop circles are interpreted as representing astronomical objects and events, such as planetary alignments, solar systems, or constellations.
    - These formations are sometimes seen as attempts to convey astronomical information that could be relevant for understanding celestial mechanics or even space travel technology.

3. **Messages About Human Technology**:
    - Some interpretations suggest that certain crop circles are warnings or messages about human technology, particularly concerning its destructive potential or misuse. This includes warnings about nuclear technology, environmental degradation, or other risks associated with advanced technology.

4. **Representation of Advanced Technologies**:
    - There are claims that some crop circles represent advanced technologies themselves, potentially hinting at methods of energy generation, propulsion systems, or other advanced engineering concepts.
    - This interpretation is highly speculative and often involves extrapolating complex technical ideas from basic geometric shapes.

5. **Binary and Digital Codes**:
    - A few crop circles have been interpreted as containing binary or digital codes. The most famous example is the 2001 Arecibo Reply formation in Chilbolton, UK, which enthusiasts claim mimicked the Arecibo message sent into space in 1974, but altered to represent an extraterrestrial source.
    - These codes are sometimes interpreted as attempts to communicate using a more universal, mathematical language that could be related to computer technology or digital communication.

The notion that some crop circles represent advanced technologies is a fascinating aspect of the broader crop circle phenomenon. Enthusiasts and theorists suggest that the intricate designs and patterns found in crop circles are not mere art, but rather sophisticated representations or blueprints of advanced technological concepts. Let's delve into more details about these claims:

Representation of Energy Generation and Propulsion Systems

1. **Energy Generation**:

    - **Designs Resembling Power Sources**: Some crop circles are interpreted as symbolic representations of advanced energy generation systems. For instance, formations resembling the structure of atoms or molecular models have led to conjectures about them being blueprints for new, possibly clean, or even free energy technologies.
    - **Geometric Patterns and Theoretical Physics**: Certain patterns are likened to complex geometrical structures found in theoretical physics, such as toroidal energy fields, which are central to concepts like zero-point energy or fusion power.

2. **Propulsion Systems**:

    - **Craft Propulsion**: Some crop circles, particularly those with radial symmetry or intricate geometric configurations, are thought to represent the propulsion systems of extraterrestrial spacecraft. Theories range from anti-gravity to more exotic forms of space travel, like warp drives or dimensional travel.
    - **Mathematical Models**: Complex mathematical models are sometimes inferred from the precise geometry of crop circles, with suggestions that these represent unknown physical principles or advanced engineering concepts relevant to propulsion.

Advanced Engineering Concepts

1. **Blueprints for Machines**:

   - **Machinery and Device Blueprints**: Some enthusiasts argue that certain crop circles could be interpreted as schematic diagrams for machines or devices of unknown purpose. These could range from simple mechanical devices to complex systems involving advanced physics.
   - **Interlocking Patterns and Mechanics**: Patterns that feature interlocking designs or repeating geometrical sequences are sometimes seen as representing mechanical systems or engineering principles not yet understood or discovered by modern science.

2. **Representation of Theoretical Models**:

   - **Quantum Mechanics and Beyond**: Some formations are believed to symbolize principles from quantum mechanics, string theory, or other advanced areas of physics. Proponents of this view argue that these crop circles might be a way of communicating complex scientific ideas through universally understandable geometric symbols.

Context of Extrapolation

- **Extrapolating from Geometry**: The process of interpreting these crop circles often involves extrapolating from basic geometric shapes and patterns to hypothesize about advanced technological concepts. This extrapolation is highly speculative and subjective, relying on the interpreter's background, imagination, and familiarity with scientific and technological concepts.
- **Symbolic Interpretation**: In many cases, the interpretations are symbolic rather than literal. A simple circular pattern, for example, might be seen as representing an energy field or a propulsion method based on how the interpreter conceptualizes these technologies.

In conclusion, the idea that crop circles represent advanced technologies is a topic of interest and extrapolation among enthusiasts. These interpretations showcase the human penchant for creativity and the desire to find deeper meanings in mysterious phenomena.

24

What are the messages, insights, and communications that human experiencers are receiving from the Extraterrestrial Biological Entities. Are they focused on the planet's future? or are there specific technologies or scientific thrusts that are being communicated?

The accounts of human abductees, often referred to as "experiencers" within the ufology community, vary widely and are deeply subjective. It's important to note that these accounts are anecdotal and can lack empirical evidence, which makes them controversial and viewed with skepticism by the mainstream scientific community. However, within the narrative accounts of these experiences, several common themes do emerge:

1. **Environmental Warnings**: A recurring theme in many abductee accounts is the concern expressed by EBEs about Earth's environmental degradation. These messages often emphasize the urgency of addressing issues like climate change, deforestation, and pollution, suggesting that these problems pose a significant threat not just to humanity but to the planet.

2. **Technological Caution**: Experiencers frequently report being cautioned about the risks associated with unbridled technological advancement. This often includes warnings about the development and proliferation of weapons of mass destruction and the potential for

technology to exacerbate environmental degradation or lead to societal harm.

3. **Spiritual Evolution and Unity**: Many abductees describe receiving messages about the importance of spiritual growth and unity among humans. These messages often stress the need for humanity to overcome its divisions and conflicts, fostering a sense of global community and cooperation.

4. **Human Potential and Capabilities**: Some experiencers report being told that humans possess untapped potential, whether in terms of mental abilities, spiritual depth, or technological innovation. These messages suggest that humanity has much more to discover about itself and its capabilities.

5. **Cosmic Perspective and Interconnectedness**: Experiencers often describe being shown Earth from space, emphasizing its fragility and the interconnected nature of all life. These experiences can impart a sense of Earth being part of a larger cosmic community, with a responsibility to act as stewards of the planet.

6. **Hybridization Programs**: In some accounts, there is a theme of EBEs engaging in genetic or biological experiments, often described as hybridization programs. The purposes speculated for these programs range from the survival of the alien species to enhancing human capabilities or preparing for future environmental conditions on Earth.

7. **Policy of Non-Interference**: Several experiencers feel that EBEs adhere to a policy of non-interference, suggesting that while they may offer guidance or warnings, they believe it is ultimately up to humans to resolve their own challenges and chart their future.

8. **Personal Transformation**: Individuals often report experiencing profound personal changes following their encounters. These can include shifts in worldview, heightened environmental consciousness,

and increased empathy and compassion, suggesting a transformative impact of the experience itself.

9. **Transmission of Advanced Knowledge**: There are claims from some experiencers of receiving advanced scientific knowledge or insights from EBEs. However, these claims often lack detailed information or specificity that would allow for empirical testing or practical application in human technology or science.

10. **Advocacy for Peace and Disarmament**: Messages advocating for global peace and the dismantling of weapons, particularly nuclear arms, are also common. These narratives often convey a sense that EBEs are concerned about humanity's potential for self-destruction and the implications of such actions for the planet and possibly beyond.

In summary, the messages and communications reported by human experiencers of extraterrestrial encounters often revolve around themes of environmental stewardship, technological responsibility, spiritual evolution, and the potential for human growth and unity. These narratives offer a compelling and often hopeful vision of what humanity could achieve and the challenges it must overcome. It's crucial to approach these accounts with a critical eye, recognizing the influence of cultural narratives, psychological factors, and the very human tendency to seek patterns and meaning, especially in extraordinary experiences. Whether these messages are genuinely from extraterrestrial entities or the product of complex psychological phenomena, they often reflect human concerns and hopes for the future.

25

Describe ten plausible events that changed or dramatically influenced the trajectory of human civilization and are linked to stories or evidence of extraterrestrial / alien or spiritual intervention/ guidance.

1. **Ancient Egyptian Civilization**: The monumental achievements of Ancient Egypt, particularly the construction of the pyramids, have long been a subject of fascination. Some theorists propose that the Egyptians had assistance from extraterrestrial beings. This theory is partly based on the pyramids' architectural complexity, their precise alignment with Orion's Belt, and the advanced mummification and medical techniques of the Egyptians, which seem too advanced for their time. This idea is bolstered by Egyptian hieroglyphs that some interpret as depicting spacecraft or alien beings.

2. **Mayan Civilization**: The Mayans, known for their sophisticated calendar and astronomical systems, have been at the center of alien intervention theories. Their Long Count calendar, which accurately predicted celestial events, has been cited as too advanced to have been developed without extraterrestrial help. Some suggest that carvings and relics from Mayan ruins depict visitors from other worlds, influencing their understanding of astronomy and time.

3. **The Nazca Lines**: In the Nazca Desert of Peru, massive geoglyphs depict various animals and shapes, visible only from high altitudes. Their purpose and construction methods are still debated. Some theorists

argue that the lines served as a kind of landing strip for alien spacecraft or as messages to beings from the stars. The precision and scale of these lines challenge explanations of how a pre-technological society could have created them without guidance.

4. **The Mahabharata and Vimanas**: In ancient Indian texts, 'vimanas' are described as flying chariots or aircraft, leading some to speculate about ancient alien visitations. These stories are rich in details that resemble modern UFO descriptions, suggesting that ancient Indians might have had encounters with advanced technology provided by extraterrestrial beings. The texts describe these vimanas using a form of energy that could be likened to modern concepts of anti-gravity or other advanced propulsion methods.

5. **The Dogon Tribe and Sirius B**: The Dogon people of Mali have a detailed understanding of the Sirius star system, including knowledge of Sirius B, a white dwarf star that is not visible to the naked eye. This knowledge predates modern astronomical discoveries, leading some to propose that the Dogon had contact with aliens from Sirius who shared this information. The specificity of their astronomical knowledge, which aligns with modern scientific understanding, remains a compelling mystery.

6. **Stonehenge**: Stonehenge's massive stones, transported from distant locations, and its astronomical alignment, particularly with the summer solstice, raise questions about the capabilities of Neolithic people. Some theorists speculate that the monument's builders had assistance from extraterrestrial visitors. This theory is supported by local legends of Merlin the magician levitating the stones, which some interpret as a metaphor for advanced technology used by aliens.

7. **Roswell UFO Incident (1947)**: The alleged crash of a UFO near Roswell, New Mexico, has become a cornerstone of UFO lore. The initial reports of a "flying disc" being recovered were quickly retracted, leading to widespread theories of a government cover-up. This incident spurred a multitude of conspiracy theories regarding alien visitations

and the government's knowledge of extraterrestrial life, profoundly impacting popular culture and UFO research.

8. **Rendlesham Forest Incident (1980)**: This incident involved U.S. Air Force personnel stationed at RAF Woodbridge in England who witnessed strange lights and an alleged landing of a craft of unknown origin in Rendlesham Forest. The detailed accounts by military personnel, including descriptions of hieroglyphic-like symbols on the craft, have led some to claim this as credible evidence of extraterrestrial visitation, influencing both public perception and the study of unidentified flying objects.

9. **The Tunguska Event (1908)**: The explosion in Tunguska, Siberia, released energy equivalent to thousands of atomic bombs, leveling millions of trees and shaking the earth. The lack of an impact crater has led to various theories, including the possibility of an exploding alien spacecraft or a test of extraterrestrial weaponry. This event remains one of the most mysterious and widely discussed examples in the study of unexplained phenomena.

10. **Crop Circles**: Crop circles, intricate patterns created by flattening crops, have been found worldwide, with some of the most complex designs appearing inexplicably overnight. While many have been created by humans, the origin of others remains a mystery. Theories range from natural phenomena to messages or signals from extraterrestrial beings. The geometric precision and complexity of certain formations exceed what many believe could be accomplished discreetly and quickly, leading to ongoing speculation about their origins.

Each of these events, whether they are based on archaeological findings, historical texts, or modern incidents, continues to intrigue, and inspire theories of extraterrestrial and spiritual intervention in human history.

26

Describe ten plausible people (inventors, visionaries, leaders) that changed or dramatically influenced the trajectory of human civilization and are linked to stories or evidence of extraterrestrial / alien or spiritual intervention/guidance.

The idea that some of history's most influential figures may have been guided or inspired by extraterrestrial or spiritual intervention is a fascinating and often controversial topic. While there is no concrete evidence to support these claims, they persist in various narratives and theories. Here are ten notable figures often linked to such stories:

1. **Nikola Tesla**: Tesla, a pioneer in electricity and magnetism, is sometimes speculated to have had contact with extraterrestrial beings. Some theories suggest that his advanced knowledge in electrical engineering and physics was imparted by alien intelligence.

2. **Leonardo da Vinci**: Da Vinci's extraordinary inventions and his ahead-of-time understanding of the human body and mechanics have led some to speculate about otherworldly inspiration. His detailed journals and visionary ideas continue to fuel these theories.

3. **Joan of Arc**: Known for her role in the Hundred Years' War, Joan claimed that her military endeavors were guided by divine voices and visions. Some interpretations of her experiences suggest a connection to extraterrestrial influence or spiritual guidance.

4. **Albert Einstein**: Einstein's groundbreaking work in physics, including the theory of relativity, has sometimes been included in theories of extraterrestrial contact. Though purely speculative, some suggest his revolutionary ideas might have been influenced by otherworldly intelligence.

5. **Socrates**: The Greek philosopher often spoke of a "daemon" or inner voice that guided his actions and thoughts. While traditionally considered a spiritual guide, some modern interpretations have linked this to potential extraterrestrial communication.

6. **Galileo Galilei**: As a revolutionary astronomer who faced significant opposition from the church, some speculative theories suggest Galileo's insights into the cosmos were influenced by extraterrestrial sources, although these claims are not supported by historical evidence.

7. **Sir Isaac Newton**: Newton's laws of motion and his work on gravity were monumental in the field of physics. Some fringe theories suggest that his profound understanding of the universe might have been influenced by external, possibly extraterrestrial, guidance.

8. **Christopher Columbus**: Some theories, albeit highly speculative, suggest that Columbus's voyage to the New World was influenced by otherworldly guidance, citing his determination to sail west despite the common beliefs of his time.

9. **Steve Jobs**: The co-founder of Apple Inc., known for his visionary approach to technology and design, has been the subject of speculation regarding the source of his innovative ideas, with some theories stretching to include extraterrestrial inspiration.

10. **Elon Musk**: As a contemporary figure pushing the boundaries of space travel, electric vehicles, and AI, Musk's ambitious visions have led to speculation about where he draws his inspiration, with some theories playfully suggesting extraterrestrial influence.

While these theories are often based on speculation and lack empirical evidence, they reflect the human fascination with the possibility of higher intelligence and guidance, whether spiritual or extraterrestrial, playing a role in the leaps of innovation and thought that have shaped civilization.

27

Who are the top ten ET contactees along with their messages for humanity?

The concept of "ET contactees" refers to individuals who claim to have had direct contact with extraterrestrial beings. These claims often include receiving messages or information intended for humanity. The stories and messages from these contactees vary widely and are generally considered speculative, lacking empirical evidence. Below are ten notable figures who have made such claims, along with a summary of their alleged messages:

1. George Adamski

- **Experience**: Meeting with beings from Venus.
- **Detailed Message:** Adamski's alien contacts emphasized the importance of global peace, warning of the destructive potential of nuclear weapons. They purportedly advocated for a more unified Earth, stressing the need for humanity to transcend national and ideological divides. The Venusians, as described by Adamski, were concerned about Earth's trajectory, particularly with respect to atomic weaponry, and sought to guide humanity towards a more harmonious and spiritually advanced state.

2. Billy Meier

- **Experience**: Ongoing contact with Pleiadians.

- **Detailed Message**: Meier's Pleiadian contacts reportedly convey deep concerns about environmental degradation and the spiritual stagnation of humanity. They emphasize the interconnectedness of all life and the importance of living in harmony with nature. These beings also speak about the need for inner spiritual development, urging humans to cultivate peace, love, and understanding both within themselves and in their relationships with others.

3. Betty and Barney Hill

- **Experience**: Abduction by aliens.
- **Detailed Message**: While the Hills' story is more about their abduction experience, it indirectly communicated themes of the unknown and the potentially vast differences between human and extraterrestrial intelligence and technology. Their story, particularly the medical examinations they reported, suggested that extraterrestrials might have a keen interest in human biology and behavior.

4. Travis Walton

- **Experience**: Alleged abduction in 1975.
- **Detailed Message**: Walton's experience, as he recounted, was traumatic and perplexing, providing more insight into the mysterious nature of the alleged abductors than a clear message for humanity. His story contributes to the narrative of alien abductions, raising questions about the intentions and interests of these purported extraterrestrial beings.

5. Whitley Strieber

- **Experience**: Recurring abduction experiences.
- **Detailed Message**: Strieber's accounts delve into the transformative impact of his encounters. He speaks of an expanded consciousness and a new understanding of reality triggered by these experiences. Strieber's narrative often blurs the line between

physical and metaphysical, suggesting that human understanding of reality is limited and that these encounters could be a gateway to a broader comprehension of the universe.

6. Elana Danaan

- **Experience**: Continuous communication with ET beings.
- **Detailed Message**: Danaan's messages from extraterrestrial beings are centered around the themes of spiritual awakening and the elevation of human consciousness. She speaks of a coming age where humans will more fully understand their place in the cosmos and develop abilities to communicate and interact with other intelligent life forms in a more meaningful way. She emphasizes the importance of humanity evolving beyond its current state of conflict and division.

7. Simon Parkes

- **Experience**: Contacts with different extraterrestrial entities.
- **Detailed Message**: Parkes discusses a complex narrative involving extraterrestrials' influence on Earth's political and spiritual realms. He suggests that humanity is at a crossroads, with the potential for significant advancement or decline, depending in part on how these extraterrestrial influences are navigated. His messages often involve the idea of a behind-the-scenes battle between different extraterrestrial factions with varying agendas for humanity.

8. Alex Collier

- **Experience**: Ongoing interactions with Andromedan beings.
- **Detailed Message**: Collier's messages from the Andromedans focus on a call for spiritual awakening and unity among humans. He warns of manipulative extraterrestrial species and emphasizes the importance of Earth humans becoming more aware of their true potential and cosmic heritage. The Andromedans, as described by

Collier, advocate for a future where humans live in harmony with each other and responsibly steward the Earth.

9. Luis Fernando Mostajo Maertens

- **Experience**: Contact with beings from Venus and the Pleiades.
- **Detailed Message**: Maertens shares messages about the necessity for a spiritual and ecological transformation on Earth. His extraterrestrial contacts purportedly stress the importance of living in balance with the natural world and foresee a future where humanity will openly interact with benevolent extraterrestrial beings. These messages often include themes of inner transformation, global peace, and the development of a more conscious and harmonious human society.

10. Ricardo González

- **Experience**: Contacts with extraterrestrial beings from a young age.
- **Detailed Message**: González speaks about messages of universal love, peace, and the necessity for a spiritual renaissance among humanity. His alleged extraterrestrial contacts encourage a shift in human consciousness towards greater empathy, unity, and understanding, both within the human race and in relation to potential extraterrestrial civilizations. The overarching theme is one of a significant evolutionary leap for humanity, moving towards a more spiritually enlightened and cosmically connected existence.

28

What are ten potential reasons that extraterrestrial biological entities (aliens) and or their craft are biological in nature rather than mechanical artificial intelligence robots and interplanetary space ships?

1. **Evolutionary Adaptation**: Extraterrestrial life forms could have evolved in diverse environments, leading to a vast array of biological adaptations. Just as life on Earth adapted to different climates and conditions, alien life might have developed unique physiological traits suited to their native environments. These adaptations could provide natural advantages in space travel, such as resistance to radiation, extreme temperatures, or the ability to hibernate during long journeys. The complexity of biological evolution might result in sophisticated, naturally occurring systems that could be more efficient and resilient than any artificially created technology.

2. **Self-Replication and Healing**: Biological systems' ability to self-replicate and heal could be crucial for extended space missions. In an environment where damage from cosmic radiation or micrometeoroids is likely, a spacecraft that can repair itself at the cellular or molecular level would be invaluable. Such a system would be able to replace or regenerate damaged parts, potentially using materials found in space, thereby greatly reducing the need for carrying spare parts or repair tools. This capability also implies that the spacecraft could adapt its structure and function in response to the changing demands of the mission.

3. **Energy Efficiency**: Biological systems, particularly those that have evolved in resource-scarce environments, may be highly efficient at energy utilization and conservation. For instance, an alien species might have developed ways to harness energy directly from stars or cosmic radiation, similar to how plants on Earth use photosynthesis. Such efficiency would be particularly advantageous in space exploration, where energy sources are limited. A biological spacecraft might also be able to store energy in bio-chemical forms more efficiently than traditional mechanical batteries.

4. **Symbiotic Relationships**: A symbiotic relationship between aliens and their crafts could lead to highly efficient and sustainable space travel. The craft, as a living entity, could be attuned to the needs of its inhabitants, providing oxygen, food (through bio-regenerative systems), and waste recycling. In return, the aliens could provide the craft with necessary nutrients and maintenance, creating a closed-loop system that reduces the need for external resources.

5. **Integration with Natural Ecosystems**: Biological entities and crafts would likely be designed to integrate with, rather than dominate, new ecosystems. This approach would be particularly important for missions involving exploration or colonization. A biological spacecraft could, for instance, be capable of harmonizing its impact on local environments, reducing ecological footprints, and even contributing positively to alien ecosystems through processes like seed dispersal or pollination.

6. **Adaptive Capabilities**: Adaptability is a key trait of biological entities. An alien species or craft that is biologically based might possess the ability to modify its physiological structure in response to new environments, such as changing atmospheric compositions, gravity, or temperatures. This flexibility could allow for exploration of a wide range of planetary types without the need for extensive reengineering or modification, as would be the case with mechanical crafts.

7. **Reduced Need for Artificial Intelligence**: If alien species and their crafts are inherently biological, they might possess natural intelligence,

awareness, or even consciousness. This innate intelligence could negate the need for developing separate AI systems for navigation, decision-making, or problem-solving during space missions. The biological craft could intuitively respond to the needs and commands of its inhabitants, creating a more integrated and responsive system.

8. **Molecular Manufacturing and Resource Utilization**: The ability of biological systems to manipulate matter at a molecular level (biomolecular manufacturing) could be highly advantageous in space. This capability would allow a spacecraft to repair itself, grow, or even change its configuration using materials available in its environment. This type of organic manufacturing and resource utilization would be invaluable for long-duration missions, where resupplying from Earth is not feasible.

9. **Biocompatibility and Reduced Harm**: Biological crafts and entities would likely be more compatible with other life forms, reducing the risk of harmful contamination in alien ecosystems. This aspect would be critical in ensuring that exploration and potential colonization do not disrupt or destroy indigenous life. The use of biologically based technologies might also be less intrusive and more harmonious with the natural environments encountered in space.

10. **Cultural or Ethical Reasons**: The preference for biological systems over mechanical or artificial ones could reflect deeper cultural or ethical values of the extraterrestrial species. They might hold beliefs that emphasize living in harmony with nature, or they might have ethical prohibitions against creating artificial intelligence. This cultural or ethical framework could lead them to develop and utilize technologies that are more aligned with natural, organic processes, viewing this approach as more ethical or spiritually fulfilling.

Each of these expanded points suggests that biological entities and crafts, as conceptualized in the context of extraterrestrial life, could offer numerous advantages in terms of adaptability, sustainability, and efficiency for space exploration and travel, shaped by evolutionary, environmental, and cultural factors.

29

What would be the impact on society and government of an artificial intelligence knowledge portal or website that is populated with the unique and rare collection of official government documents from the national archives along with curated UFO related books, periodicals, and conference proceedings, comprising millions of pages of testimony, insights, and reports about the phenomenon. Could this facilitate or accelerate the motivation for governments to admit the ET reality?

The creation of an extensive and publicly accessible artificial intelligence knowledge portal containing official government documents, books, periodicals, and conference proceedings related to UFO phenomena could have a significant impact on society and government in several ways:

1. **Increased Public Awareness and Education**: Such a portal would serve as a central repository for information, making it easier for the public to access and learn about UFO phenomena. This could lead to a more informed public discourse, moving the topic from fringe to mainstream discussions. Educational institutions might also incorporate this data into curricula, further legitimizing the study of UFO phenomena.

2. **Advanced Data Analysis Capabilities**: The integration of AI could revolutionize the way UFO-related data is analyzed. AI algorithms could detect subtle patterns or connections across millions of documents, potentially uncovering new insights or validating existing

theories. This could lead to breakthroughs in understanding the nature, frequency, and implications of UFO sightings and encounters.

3. **Transparency and Government Accountability**: The portal's existence could act as a catalyst for greater government transparency. As the public gains access to previously classified or hard-to-find documents, it could lead to increased scrutiny of government actions and policies regarding UFOs, potentially fostering a more open dialogue between the public and government entities.

4. **Influencing Public Opinion and Policy**: With a wealth of information readily available, public opinion on UFOs and extraterrestrial life could shift significantly. This, in turn, could influence policy decisions, leading to more funding for research, the establishment of dedicated governmental bodies, or the development of protocols for potential extraterrestrial contact.

5. **Stimulating Global Scientific and Academic Engagement**: The portal could legitimize the study of UFOs within the scientific and academic communities. Researchers across various disciplines – from astrophysics to anthropology – might engage more deeply with the material, leading to interdisciplinary studies and potentially groundbreaking scientific papers.

6. **Cultural and Philosophical Implications**: As UFO phenomena become part of the mainstream conversation, this could have significant cultural and philosophical impacts. It might alter our collective understanding of human history, our place in the cosmos, and the nature of reality itself.

7. **Debunking or Validating Conspiracy Theories**: The portal could either debunk widely held conspiracy theories by providing factual information or lend credence to some theories if the data supports them. This could lead to a reevaluation of historical events and narratives.

8. **Encouraging International Cooperation**: The global nature of UFO phenomena might encourage international cooperation in sharing information, research, and strategies for potential extraterrestrial engagement, fostering a sense of global unity in facing a common, albeit unknown, subject.

9. **Potential for Misinformation and Misinterpretation**: While AI can efficiently process and analyze data, there's a risk of misinformation or misinterpretation, especially if the AI's algorithms are not transparent or if the data is taken out of context. This could lead to public confusion or the spread of unfounded theories.

10. **Impact on Religious and Spiritual Beliefs**: The portal could also impact religious and spiritual beliefs. For some, it might reinforce a belief in a higher power or interconnected universe, while for others, it could challenge fundamental religious doctrines.

11. **Influence on Media and Entertainment**: The entertainment industry might draw inspiration from the portal, influencing movies, television, books, and other media. This could further embed UFO and extraterrestrial themes into popular culture.

12. **Economic Implications**: The portal could spur new economic opportunities, from the development of related technologies to tourism in areas known for UFO activity. It could also impact investment in space exploration and related industries.

In summary, an AI-enhanced UFO knowledge portal could be a pivotal tool in transforming public understanding and government transparency regarding UFO phenomena. Its impact would be wide-ranging, affecting everything from scientific research and policymaking to cultural norms and global collaboration. However, the effectiveness of such a portal in achieving these outcomes would heavily depend on the integrity of the data, the sophistication of the AI analysis, and the public's willingness to engage with and interpret the information presented.

While the portal could facilitate a greater understanding and potentially motivate governments toward more transparency, it's also possible that without clear and incontrovertible evidence of extraterrestrial reality, governments may continue to be cautious in their official positions. The impact would also be shaped by the societal context, media treatment, and the prevailing public attitudes towards government transparency and the possibility of extraterrestrial life.

30

What are the five main forces that are interconnected driving global challenges? How are government debt and money economic forces affecting this? How is the political and societal polarization of the left and right mindset and actions influencing our future? Compare and contrast the geopolitical attitudes, policies, and priorities of countries particularly US and China when it comes to the planets future evolution? Addressing climate changes and mitigation costs will be a major global expense, how should equitable cost sharing be devised? Technology progress in energy, artificial intelligence, and biological sciences such as CRISPR, are they key enablers to a more peaceful planet with less war, inhumanity, and poverty?

Five Main Forces Driving Global Challenges

1. **Technological Advancements**: Rapid technological progress, particularly in AI, biotechnology, and digital communication, is reshaping economies, societies, and individual lives. This force can be a double-edged sword, offering solutions (e.g., in healthcare, energy efficiency) while creating new challenges (e.g., job displacement due to automation, privacy concerns).

2. **Environmental and Climate Change**: The escalating impact of climate change, loss of biodiversity, and environmental degradation poses a significant threat to global stability and well-being. These challenges require coordinated global action but often face local resistance due to immediate economic costs.

3. **Economic Inequality**: Growing economic disparities both within and between countries contribute to social unrest and hinder collective action on global issues. The concentration of wealth and the widening gap between the rich and the poor can lead to a cycle of poverty and social inequality.

4. **Political and Ideological Polarization**: Increasing polarization, both within countries (like the U.S.) and globally, hampers effective governance and international cooperation. This polarization is often fueled by misinformation, social media echo chambers, and nationalist sentiments.

5. **Global Health Challenges**: Pandemics like COVID-19 demonstrate the interconnectedness of global health and the need for cooperative approaches to health crises. Emerging health threats, often exacerbated by climate change and global mobility, require robust, coordinated responses.

Government Debt and Economic Forces

Government debt and economic policies play a crucial role in addressing global challenges. High levels of debt can limit a government's ability to invest in essential areas like healthcare, education, and climate change mitigation. However, strategic debt can also be a tool for stimulating economies and funding critical initiatives. Economic forces, including trade policies, currency fluctuations, and financial markets, significantly influence global stability and development. The challenge lies in balancing fiscal responsibility with necessary investments in sustainable development.

Political and Societal Polarization

The polarization between 'left' and 'right' ideologies significantly impact our future. It often leads to gridlock in policymaking, reducing the effectiveness of governments to address critical issues like climate change, social inequality, and technological regulation. This division can also fuel social unrest and hinder international cooperation, as countries become more inward-looking and less willing to participate in global solutions. Overcoming this polarization is essential for a collaborative approach to global challenges.

Geopolitical Attitudes: US vs. China

- **United States**: Traditionally, the U.S. has played a leading role in global affairs, advocating for democratic values, free markets, and international cooperation (e.g., in climate agreements, NATO). However, recent years have seen a shift towards nationalism and a focus on 'America First' policies.
- **China**: China's approach is more state-centric, focusing on economic growth, technological advancement, and increasing its influence in global affairs. China's Belt and Road Initiative and its moves in the South China Sea illustrate its expanding geopolitical ambitions.

Both countries have different views on governance, human rights, and global cooperation, impacting global policymaking, especially in areas like climate change, trade, and technological standards.

Addressing Climate Change: Equitable Cost Sharing

Equitable cost sharing for climate change mitigation is complex. Developed countries, having contributed more historically to greenhouse gas emissions, have a greater responsibility and capacity to bear costs. However, emerging economies argue for their right to economic development. A fair approach might include:

- **Scaled Contributions**: Based on GDP, historical emissions, and current emission levels.

- **Technology Transfer**: Developed countries providing technology and support to developing nations.
- **Green Financing**: Global funding mechanisms to support climate initiatives in poorer nations.

Technology as a Key Enabler for Peace

Advancements in energy, AI, and biotechnologies like CRISPR hold immense potential for addressing global challenges:

- **Energy**: Renewable energy technologies can reduce dependence on fossil fuels, mitigating climate change and reducing resource-related conflicts.
- **Artificial Intelligence**: AI can enhance efficiency in various sectors, aid in solving complex global issues, and improve living standards. However, it also poses risks like autonomous weapons and surveillance.
- **Biological Sciences**: CRISPR and similar technologies can revolutionize healthcare, agriculture, and environmental conservation, potentially reducing disease and food scarcity.

However, these technologies must be governed ethically and equitably to ensure they contribute to peace and do not exacerbate inequalities or create new forms of conflict.

Integration with UFOs/UAPs and ET Presence

The belief in or potential disclosure of extraterrestrial presence (UFOs/UAPs) could significantly impact these areas:

- **Technological Advancements**: Recovered extraterrestrial technology could leapfrog current capabilities, especially in energy and propulsion, potentially offering solutions to environmental challenges.
- **Global Unity**: The confirmed existence of extraterrestrial life could unify humanity under a shared global identity, potentially reducing ideological polarization.

- **Geopolitical Shifts**: The revelation of ET presence might prompt major powers like the US and China to cooperate on a global response, altering traditional geopolitical dynamics.
- **Ethical and Philosophical Questions**: Such a disclosure would raise profound questions about humanity's place in the universe, potentially impacting religious beliefs and philosophical worldviews.

In conclusion, addressing these interconnected global challenges requires a multifaceted approach, balancing technological innovation with ethical governance, economic equity, and international cooperation. The potential influence of extraterrestrial realities adds another layer of complexity and opportunity to this global endeavor.

Made in United States
Orlando, FL
11 November 2026

90609392R00072